HEAVY WEATHER GUIDE

HEAVY WEATHER GUIDE

PART I—HURRICANES

by Captain Edwin T. Harding, U. S. Navy

PART II—TYPHOONS

by Captain William J. Kotsch, U. S. Navy

United States Naval Institute, Annapolis, Maryland

Library of Congress Catalogue Card No. 64-25388

Printed in the United States of America
By George Banta Co., Inc.
Menasha, Wisconsin

Price—$6.00

Publisher's Preface

Sweeping up from tropical zones into the colder latitudes of the northern hemisphere, vast cyclonic storms each year leave in their wakes death and devastation. In the Atlantic these storms are called hurricanes; in the Pacific they are called typhoons. Both kinds of storms carry inconceivable amounts of energy and have greater potential for destruction than the most powerful atomic bombs. They are mysterious, yet they follow generally predictable behavior patterns; they are terrifying, yet their approach need not lead to panic. The purpose of this book is to discuss the theory, origin, and habits of these storms, and to show how to evade or survive them.

This book treats hurricanes and typhoons separately, and so there is some repetition, because in most ways hurricanes and typhoons are alike. Yet due to the far-ranging scale of operations of both naval and merchant vessels, both kinds of storms are included in one book, for the skipper who is concerned about a hurricane off Bermuda this month may well face the prospect of a typhoon off Guam a few weeks hence.

The reader, whether he be deep-draft skipper, weekend sailor, or shore-side dweller, can learn about hurricanes and typhoons from this book without actually experiencing one. For those who wish the vicarious experience of weathering one of these awesome storms, there are included excerpts from *Typhoon,* by Joseph Conrad, as well as other accounts of tropical storms by seamen who have experienced them.

Table of Contents

APPENDIXES

INDEXES

List of Figures

PART I—HURRICANES

PART II—TYPHOONS

HEAVY WEATHER GUIDE

PART I—HURRICANES

Section 1—The Tropics

Section 2—The Hurricane

Section 3—Characteristics of the Hurricane

Section 4—The Killer Side of Hurricanes

Section 5—The Hurricane Warning Service

Section 6—Tips from the Old Weatherman

Section 7—The Future

Introduction

In 1943, our small island base was hit by a tropical cyclone. As soon as the winds died, the Captain ordered a cleanup job. This was my first hurricane, but I knew the worst was yet to come, and tried to convince him of further danger. We were merely in the calm eye of the storm; the second half would bring winds as strong as in the first half, and from a more dangerous direction. If my skipper had any doubts about this, they vanished when the storm hit again half an hour later.

Since 1943, in firsthand experience with ten tropical cyclones and in forecasting on over a hundred, I have observed the general degree of uncertainty people display in making related operational decisions, an uncertainty bred by lack of knowledge. This uncertainty may be that of the ship's captain trying to decide whether he should outrace the storm or heave to and watch it go by; or of the owner of a large building who must decide whether he should spend thousands of dollars boarding it up; or of the admiral who must decide whether to send all ships and aircraft out of a threatened base, or in wartime, whether or not to disperse a whole fleet, thereby jeopardizing the individual units or possibly a planned strike.

Decisions relating to hurricanes must be based on a full understanding of hurricane life cycles and characteristics. It would seem that the necessary knowledge should be widespread enough in areas affected by hurricanes, considering the number of people who are involved over the span of a few hurricane seasons and the publicity generated by each storm. But circumstances militate against general knowledge. There are relatively few hurricanes per year, so our experience is limited as compared to the frequent contacts with mid-latitude storms which so often make fine forecasters out of sailors and farmers; also, many new residents move into hurricane-affected areas each year. Books and articles on hurricanes are usually too technical or too long for most to read. Popular books, movies, and TV programs with hurricanes as background present a distorted picture of the hurricane, because they are primarily concerned with the emotional reactions of a hero and a heroine—the latter not only hysterical but seeming never to wear hurricane-proof clothes.

This guide is short and non-technical; yet, it provides all the hurricane knowledge basic to decision making for those at sea in ships or boats, for personnel at shore installations, and for people in cities or in exposed beach areas.

EDWIN T. HARDING
Captain, U. S. Navy

Washington, D.C.
18 January 1965

Section 1 : The Tropics

Stage Setting for the Hurricane

Before the entrance of the star performer—the hurricane—it is necessary to introduce other players, and this can be done with key words and concepts related to weather in the tropics. These words and concepts describe the framework within which the hurricane is born, lives, and often dies.

The story of the tropics can be much more authoritative today than it could have been in the early years of World War II, when it became evident how little was known about tropical meteorology. Most weathermen then had been trained in the Norwegian mid-latitude theories. Their thinking, regardless of the latitude of the area for which they were forecasting, was in terms of cold and warm fronts and the requirement for ice crystals in cloud tops to produce rain. After handing out advice a few times in the tropics on where to hide ships or naval task forces under frontal clouds that did not develop, or having to abort target strikes because of rain from clouds whose tops were at a temperature ten to fifteen degrees above the freezing level supposedly necessary for rain, military weathermen generally wound up in a completely baffled state. There was a hurried effort to start research on the tropical weather problem, and weathermen learned much of value during the last half of the war. Since the war, they have continued the research and have gathered more knowledge about the tropics in the past two decades than the studies of previous centuries had accomplished. Much of the new knowledge is of use in making operational decisions.

The Bermuda High and the Trade Winds

The hurricane performs on a broad stage—the Atlantic Ocean, the Caribbean Sea, and the Gulf of Mexico. At the approximate center of action, as shown in Figure 1, is the Bermuda High, or Bermuda-Azores Area of High Pressure. The tropical section of this area, from about 30° North to the Equator, controls the whole tropical weather scene, the usual and the unusual. It is the birthplace and home (for all but the dying stages of some) of all Atlantic hurricanes.

The high center fluctuates north and south a few degrees at a time during the hurricane season, with varying degrees of intensity during these movements. Persistent departures from the normal position and strength have marked effect upon hurricane frequency and path.

North of the Bermuda-Azores High are the prevailing westerlies and to the south are the northeast and east trade winds, which extend all the way from the Spanish and African coasts to the East and Gulf Coasts of the United States. The trade winds are the most persistent in direction of any "wind belt," an advantage used unwittingly but successfully by the early sailing ships. The first voyage of Columbus is an excellent example. As Colum-

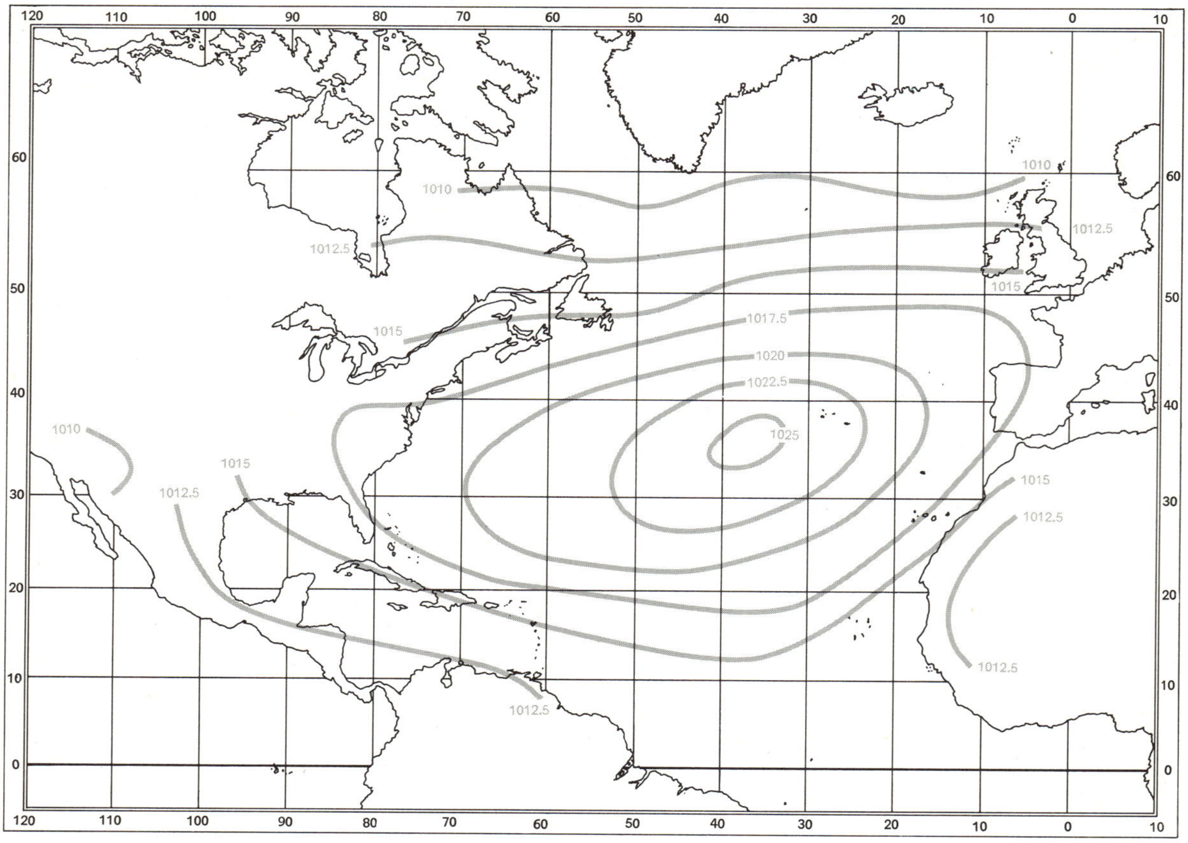

Fig. 1 Mean sea level pressure (millibars) over the North Atlantic in August.

Fig. 2 Prevailing winds over the oceans during August. Scale indicates relative constancy of the wind.

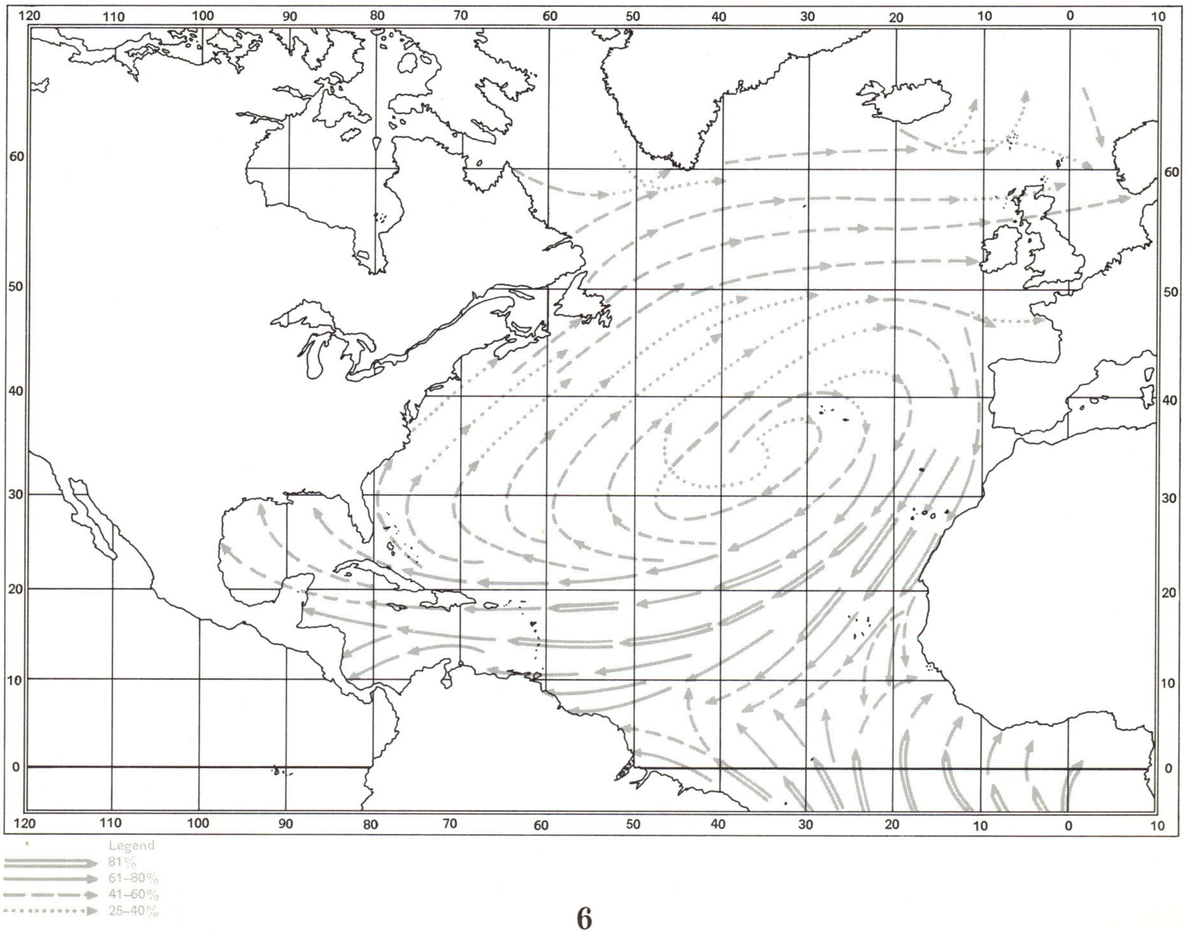

bus weighed anchor and sailed westward from the Canaries, his ships were immediately caught in strong, northeast winds. For ten days the winds persisted; so steady and strong were they that his sailors grew more and more fearful that they would never succeed in returning home against such inexorable winds. An interlude of light winds in the Sargasso Sea reassured them, and they applied themselves more willingly to working the ship forward on a course more south of west until they caught the trade winds again and sped on their way to the New World.

Air in the lower layers of the northeast trades is characterized by warm temperatures, high humidities, and predominantly partly cloudy skies—in other words, just what you expect to find when you set off on a voyage through the tropics. But apart from these average conditions, trouble starts.

Equatorial Trough or Intertropical Convergence Zone

The prevailing wind charts for the month of August, Figure 2, show that the Northern Hemisphere northeast trades and the Southern Hemisphere southeast trades converge along a line near the Equator. The convergence zone is a permanent feature of the general circulation of the earth's atmosphere. The zone follows the sun and reaches farthest north in summer (about 20° at some points in the Atlantic); then, it moves near and occasionally south of the Equator in winter.

The area of convergence where the air from different hemispheres meets has been called by many different names: Intertropical Front, Intertropical Convergence Zone, Equatorial Convergence Zone, Equatorial Front, Equatorial Trough, Doldrums, Thermal Equator, and Meteorological Equator. The area is not really a front, as we know fronts, and convergence occurs only in parts of the zone at any one time. But there is a round-the-world trough of low pressure, so the term Equatorial Trough is most appropriate.

Cloudiness and precipitation are typical of the Equatorial Trough, but intensity varies considerably from one area to another. It all depends on the convergence —how much air from the opposing streams is being forced together. Sometimes, there will be great masses of clouds in bands 50 to 100 miles wide, with cloud heights to 40,000 feet or more. Cruising beneath or flying through, it would be easy to get the impression that you were in a front, or even in a tropical storm. At other times in the Equatorial Trough, there will be nothing but fair weather. The pattern can change rapidly, with both time and distance.

Many early forecasters, even some through World War II, believed that tropical cyclones formed principally in the Equatorial Trough or Doldrums. However, careful climatological studies and data gathered by reconnaissance aircraft in the past twenty years have shown that only rarely do incipient storm circulations form in the Equatorial Trough, particularly when it is near the Equator. This is true regardless of the intensity of the trough, so there is less danger from tropical cyclones in the Doldrums or Equatorial Trough than 5° to 10° to the north.

Easterly Waves

Easterly waves are troughs of low pressure or oscillations embedded in the northeast trades south of the Bermuda-Azores high-pressure area. They move from east to west at an average speed of 13 knots and, measuring along the direction of motion, will average 900 miles from the crest of one wave to the next crest. A station in the Caribbean will have about two wave passages a week during the hurricane season.

In 1940, Gordon Dunn, Chief of the

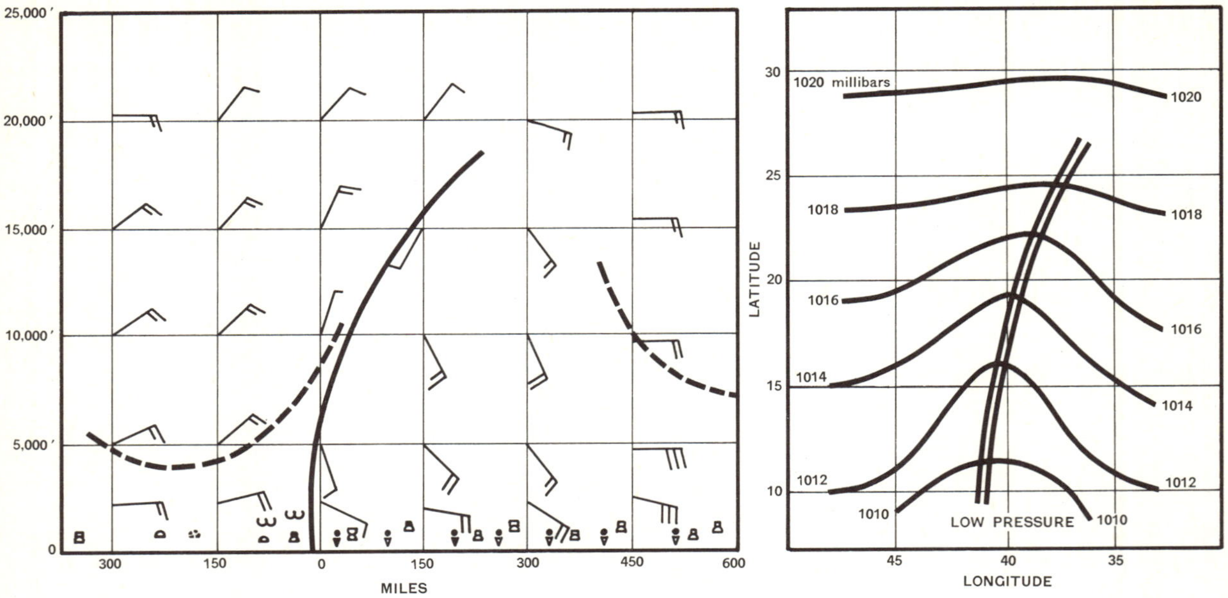

Fig. 3 (Left) *Model of an easterly wave. Depth of the moist layer is indicated by the broken line. The horizontal distances entered are measured from the position of the wave crest, and the time scale gives the average amount of time needed for a wave to pass a particular station. (After Riehl.) (From* Atlantic Hurricanes, *by Gordon E. Dunn and Banner I. Miller. Reprinted by permission of Louisiana State University Press.)* (Right) *Easterly wave on a surface weather chart.*

Fig. 4 *Latitudinal profiles of rainfall.* (Left) *December-February and June-August.* (Center) *December-February, oceans and continents separate.* (Right) *June-August, oceans and continents separate. (After Riehl.)*

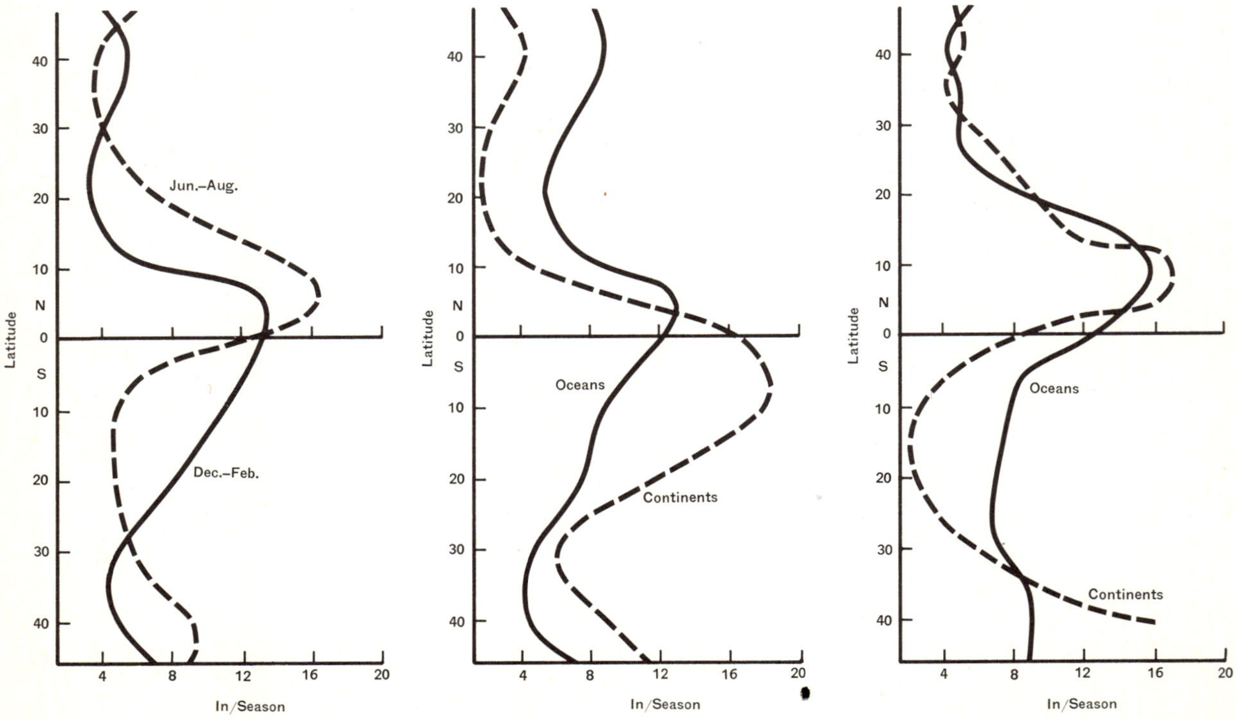

National Hurricane Center in Miami, first pointed out the regular procession of easterly waves in the Caribbean and linked their existence to the formation of hurricanes.

Figure 3 shows the typical vertical and horizontal structure of a stable easterly wave. To the west of the wave aloft, the air flows downward, and this results in good weather; at the surface, the normal east or northeast trade wind backs to northeast or north, and the pressure falls. As the wave trough approaches, the layer of moist air becomes deeper and tall cumulus clouds start to develop. To the east of the wave, the air is rising and the moist layer has maximum thickness. Large cumulus and cumulonimbus (thunderstorm) clouds become numerous, and middle- and high-cloud decks develop. Heaviest weather activity occurs 200 to 300 miles behind the trough.

The foregoing description applies to a stable easterly wave, which may travel all the way from Africa to Florida with little or no change in structure. There are also unstable easterly waves which spawn the vortices that may become hurricanes. These waves, reversing from the stable wave pattern, slope from east to west with height. Here, the bad weather is ahead of the surface low-pressure line. Reconnaissance aircraft keep close daily watch on the full length of all unstable easterly waves so as to catch future hurricanes in the embryo stage.

Rainfall

There are probably more popular misconceptions concerning rainfall than any other weather phenomenon in the tropics. For instance, newcomers to the tropics are often surprised to discover that regions around the equatorial belt with hundreds of inches of rain per year alternate with very dry regions and that there are deserts in the subtropics. Furthermore, if the time-honored legend that watches can be set by the regularity of the afternoon shower were followed, it would soon lead to some peculiar living schedules.

Figure 4 shows the latitudinal distribution of rainfall in the winter and summer seasons. There is more rainfall in equatorial regions than at high latitudes; yet, the distribution is far from uniform longitudinally. Going northward, you will note the extreme dryness over the continents between 10° to 25° in the winter seasons. This is a real desert condition, and some localities normally experience many months with no measurable rainfall. In general, analysis of rainfall records shows that no simple relationship between latitude and rainfall can be derived; there is too much variation in local conditions. At any locality in one season there may be floods, followed by drought at the same season the following year.

Turning to daily rainfall patterns, again there is wide variation. It has been the custom to describe tropical rainfall patterns on the basis of randomly distributed but monotonously regular showers. Actually, studies show that most of the tropical rainfall arises from organized weather disturbances, which occur infrequently and are characterized by narrow bands of activity. So, instead of the daily shower contribution making a rainfall pattern, 10 to 15 per cent of the rainy days accounts for as much as 50 per cent of the total rainfall.

In planning operations, travel, or a lazy tropical vacation, then, you must cast aside ideas of regularity in rainfall patterns. The pleasant warm days freshened by regular daily showers you may have been led to expect may well be days of no rain at all, or days of repeated heavy shower activity. There is much more seasonal and geographical variation in rainfall patterns in the tropics than in mid-latitudes.

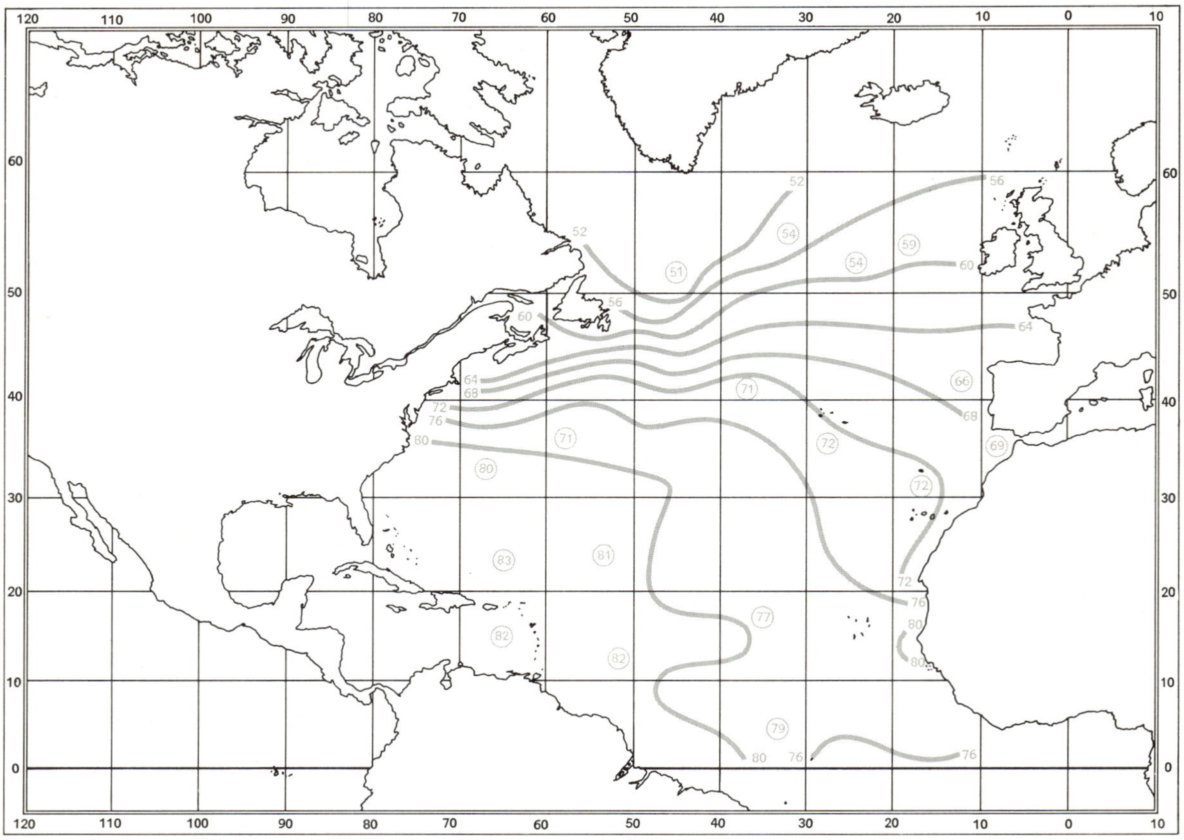

Fig. 5 Mean air temperatures and water temperatures (circled numerals) over the North Atlantic in August.

Temperatures

There may not be monotony in all phases of tropical climates, but there surely is in temperatures. The mean annual temperature range is very small; all through the Caribbean island chain the range is 10° or less, and approaches 0° as we get near the Equator. Curiously enough, inhabitants of those areas where there may be a change of but one or two degrees in the mean temperature from one month to the next are as sensitive to this as residents of the mid-Atlantic states are to the change from October to November.

Daily temperature ranges over tropical waters are correspondingly small. For instance, over water areas of the latitude of Puerto Rico, there may be daily temperature ranges of about 10°F. Daily temperature ranges increase markedly whenever land masses enter the picture.

In Figure 5, the mean air temperatures during the month of August over the North Atlantic are shown; water temperatures are nearly always within one degree of air temperatures. You will note the gradual warming trend westward from Africa to the Gulf of Mexico. (Mean air temperature patterns do not apply to waters immediately adjacent to coasts.)

Air and water temperatures are factors of critical importance to the formation and continued life of hurricanes.

Diurnal Effects

In addition to the major seasonal and annual features of tropical winds, temperatures, and rainfall, there are some daily effects with which the tropical traveler and decision-maker should be familiar.

At sea in the tropics, little variation in mean wind velocity occurs except for a slight increase of 2 to 4 knots in some areas in the period between noon and midnight. The main diurnal wind variation appears with the land and sea breezes. As land areas heat during the day, the air is warmed and rises, and pressure falls. This creates a pressure gradient from sea to land, and the air over the nearby sea moves inland, causing the sea breeze. It can be 10 to 20 knots in strength, but is not related to tropical cyclone formation. At night, the land cools and the process reverses, thereby creating the land breeze. Unless reinforced by cool air rolling down steep mountain slopes, land breezes are much weaker than sea breezes.

The small diurnal variation of temperature in the tropics at sea has already been pointed out. In zones where hurricanes form, the range will usually be 10° or less. Marked variations are usually associated with heavy cloudiness or precipitation.

As to rainfall and cloudiness, the daily pattern is for these to reach their peak inland during the day, with clear nights. The reverse is true immediately offshore. Farther out to sea, there is some increase in low cloudiness at night, and cloudiness occurs near sunrise. However, this diurnal variation is only in the range of 10 per cent of total cloudiness, not enough to cause worry about an unannounced tropical storm.

One of the most regular diurnal changes occurs in pressure. The range is from 1 to 1.5 millibars (0.0295 to 0.443 inches), and more at some places. Maxima occur at 10:00 A.M. and 10:00 P.M. local time. Minima occur at 4:00 A.M. and 4:00 P.M. The uninitiated need not be alarmed by a falling barometer unless it exceeds the diurnal range.

Section 2 : The Hurricane

Why do hurricanes acquire personality? Mid-latitude storms roar by all winter, but only one in a year or two has a personality to be remembered for more than a month.

Hurricanes acquire personality for several reasons. When a hurricane threatens any populated area, or a ship is caught in one at sea, the wire services are hot with the news. Every reconnaissance flight is reported; every movement, every quirk of the storm is discussed publicly. Neighbors gather over the back fence to talk about old-time storms, how this one compares, and what should be done about it. People in hurricane-affected areas realize the danger involved, particularly if the new hurricane has already pulverized one populated area or if its characteristics resemble those of a bad storm which occurred several years ago.

Nameless things lack personality, but now hurricanes are given girls' names. The prosaic practice of calling hurricanes "ABLE," "BAKER," etc., or using just latitude-longitude description, was cumbersome, confusing, and lacking in glamor. Then, George Stewart, in his novel *Storm,* called the storm "Maria" and started something. Girls' names as storm identifiers are short, easy to recognize, pronounce, and remember. Furthermore, they are used in alphabetical succession, so there is little chance for confusion from one storm to the next, or at a time when two or even three storms might be cavorting about the Atlantic. See Figure 6. In passing, it must be agreed that there may be more merit to the idea and practice of an early Australian forecaster who named hurricanes after political figures he disliked. The storms usually obliged with behavior which lent itself to appropriate description in the forecasts.

One last facet of the hurricane's personality is that it often demonstrates a fickle, devil-may-care attitude toward obeying any rules of movement. It steams up to a shore line; everyone buttons up and runs for cover. Then, as though deliberately taunting this show of fear, the hurricane may stop, back up, make loops, or never come back to this point again. Everyone goes crazy, particularly the hurricane forecasters. Underlying the whole problem, providing foundation for the storm's personality, is the way in which the drama unfolds. Unlike a tornado, where you see it, and then a few minutes later no one sees you, the hurricane moves ponderously, closer and closer. Fear and tension mount by the hour; when the storm really hits, the emotions are drained; you wait dully for the blow to fall.

A hurricane is a cyclone, which is an atmospheric system where the barometric pressure steadily decreases from the periphery to a minimum at the center, and where the winds spiral inward from all sides (counterclockwise in the Northern Hemisphere). The cyclone family includes systems of such widely differing character as tornadoes, temperate zone storms or lows, and hurricanes.

TROPICAL CYCLONE NAMES FOR USE IN THE ATLANTIC OCEAN, CARIBBEAN SEA, AND GULF OF MEXICO

1965	1966	1967	1968
Anna	Alma	Arlene	Abby
Betsy	Becky	Beulah	Brenda
Carol	Celia	Cindy	Cleo
Debbie	Dorothy	Debra	Dora
Elena	Ella	Edith	Ethel
Frances	Flossy	Frieda	Florence
Gerda	Greta	Ginny	Gladys
Holly	Hallie	Helena	Hilda
Inga	Inez	Irene	Isbell
Jenny	Judith	Janice	Janet
Kara	Kendra	Kristy	Katy
Laurie	Lois	Laura	Lila
Martha	Marsha	Margo	Molly
Netty	Noreen	Nona	Nita
Orva	Orpha	Orchid	Odette
Peggy	Patty	Portia	Paula
Rhoda	Rena	Rachel	Roxie
Sadie	Sherry	Sandra	Stella
Tanya	Thora	Terese	Trudy
Virgy	Vicky	Verna	Vesta
Wenda	Wilna	Wallis	Winny

Fig. 6 Tropical cyclone names for use in the Atlantic Ocean, Caribbean Sea, and Gulf of Mexico. Names are repeated every fourth year. However, names of particularly severe storms, such as "Flora" in 1963, are removed from the list. Hurricanes of this nature are subject to research, and to avoid confusion the same name is not used for later storms.

The hurricane is a beast of extraordinary violence. It does not involve as large an area as many temperate zone storms do, nor does it have the narrowly concentrated, irresistible force of the tornado. But, approximating the size of one and the intensity of the other, it is the most dangerous and destructive of all Atlantic storms. In an active hurricane year, damage may exceed two billion dollars with hundreds, perhaps thousands, of lives lost and homes destroyed.

A hurricane is defined technically as a storm of tropical origin with a cyclonic circulation reaching a strength of 64 knots (75 mph) or more at the center. In tropical warnings, by agreement in the Atlantic area, the following definitions of tropical cyclones are used:

Tropical Depression

Maximum sustained winds of 33 knots or less, and with one or more closed surface isobars.

Tropical Storm

Closed isobars and wind force between 34 and 63 knots.

Hurricane

Wind force 64 knots or more.

The word "hurricane" comes from the Spanish *huracan,* a term acquired from the Caribbean Indian tribes. There are several variations of the word among the tribes, most with a spiritual connotation. The *huracan* meant evil spirit; *hurakan* was the thunder and lightning god; *hyoracan* was a devil.

It should be remembered that the hurricane is exactly the same atmospheric phenomenon as the Pacific typhoon, the Indian Ocean cyclone, the *baguio* of the Philippines, and the willy-willy of Australia.

Formation

The tropical areas of the world are poor storm-breeding stock, and produce only about fifty tropical cyclones a year in the Northern Hemisphere. How different our history might have been if tropical disturbances were spawned as prolifically as their non-tropical cousins. It is not unusual outside the tropics to have twenty to thirty cyclones scattered around the world every day.

The relative rarity of hurricane and typhoon formation has always been, and still is, a nagging mystery to meteorologists. It nags the more because the problem seems much simpler than that of complicated, mid-latitude storms. The hurricane is the nearest thing to a simple heat engine among large-scale atmospheric disturbances, so it should be easy to form and easy to explain. Day after day, atmospheric conditions in the tropics seem conducive to storm birth—the temperatures are right, there is ample moisture in the air, the winds are neither too strong nor too weak, there are disturbances (easterly waves) in the easterly wind current which seem to have every chance for normal pregnancy and ultimate delivery as squalling hurricanes. Yet, only about once in ten times do tropical cyclones develop from these fertile looking patterns and, among this ten per cent, the development occurs in very different ways.

Sometimes, hurricanes start with a tight core of strong winds and build from the inside out; at other times, they have a large eye, strongest winds and weather well out from the center, and build from the outside in. Sometimes, there is a normal wind distribution in lower levels, but cloud structure up only to 5,000 to 10,000 feet. And such are the mysteries of hurricane formation that sometimes an aircraft flying at 35,000 feet finds everything to fit the accepted hurricane model—clockwise winds at flight level and the appropriate clouds, counterclockwise winds at lower levels—while at the surface ships are sailing peacefully on their way with no thoughts of danger. The time-frame for the formation process ranges from 6 to 12 hours up to several days.

The principal handicap in solving meteorological problems in the tropics is the lack of reliable data. Island weather stations are few in number, ship reports comparatively fewer. There is but a handful of ships and stations to provide upper-air reports; the daily average of reports between 20° to 70° N is only twenty-five. In recent years, reconnaissance and research aircraft have searched out the storms at all levels. Their information has added much to the understanding of tropical cyclones, but it has caused a degree of consternation as well by exploding some of the simplified structure theories with which meteorologists have comforted themselves over the years.

Theories of cyclone formation in the tropics are numerous, ranging from the simplest to the most complex. The earliest conception was that the storms resulted from the whim of the gods, a theory still not completely discounted by forecasters. The most popular theory of modern times was that the energy from the release of latent heat in clouds surrounding the eye started and maintained hurricanes. By some process never satisfactorily explained, huge cloud masses were supposed to get together and release the necessary energy. The rotation of the earth would start the whole thing turning in a circular motion.

More recent models of hurricane formation have continued to grow in complexity, and seem to get nearer and nearer to a satisfactory explanation. Dr. Herbert Riehl (Colorado State University) has contributed more than anyone to theoretical understanding of the formation problem, but in a recent summary in *Science* (On the Origin and Possible Modification of Hurricanes) he admits that the explanation of hurricane formation is still incomplete. It is to be hoped that future research may answer this question more definitively.

Signs

Since hurricanes are such perverse creatures and laboratory-caliber observations are never available in the right place at the right time, it has not been possible to put numbers on all the meteorological elements involved in their formation. Some definitive work has been done on sea surface temperatures; other elements must remain in the generality class until further data are available.

These are the signs which tell you that conditions are favorable, or meet necessary limits, for tropical cyclone formation:

1. Sea surface temperatures 79° or higher. (This seems to be the only critical limiting condition which is well founded at present.) Hurricanes and tropical storms will tend to dissipate in travel over waters whose surface temperatures are below this.
2. Sea level air temperature should approximate but be slightly below sea surface temperature.
3. Existence of some sort of disturbance in the tropics, and movement of this disturbance at 12 knots or less. This will customarily be an easterly wave.
4. Below normal pressure at low latitudes, above normal at high latitudes. On the low side, this would be five millibars below normal, or about 1008 mb. (29.77 inches).
5. Heavy rain or showers in the area, but these must be associated with easterly waves since most occurrences of heavy rainfall in the tropics are not associated with hurricane formation.
6. Dew point temperatures above normal, 70° or more.

The remaining indications are more in the nature of warning signs of actual tropical cyclone existence rather than just favorable conditions:

7. Twenty-four-hour sea level pressure changes exceeding the normal 2-millibar (0.06-inch) range. If the pressure fall reaches or exceeds 3.4 mb., watch out!

8. An increase in wind speed of 25 per cent or more in a limited area in the normal trade wind flow, especially if the wind flow changes cyclonically, say from easterly to a more northerly direction.

9. Any wind south through west to north of 10 knots or more in the normal easterly trade wind belt is a danger signal.

10. Thick high clouds and steady-type rainfall, rather than showery type, are fairly definite evidence that a storm has formed.

Frequency

Figure 7 shows the frequency of tropical storm origin by months within 5° squares for the years 1899 to 1958. Figures 8A and 8B show the points at which the tropical storms attained hurricane intensity (64 knots). These three figures show where hurricanes start and mature, and one can reckon his chances of encountering a storm or hurricane in any given month along any route or at any location.

It should be noted here that hurricanes do not occur in the South Atlantic. The Equatorial Trough rarely goes south of the Equator and, for comparable latitudes to the north, the sea surface temperatures are much colder south of the Equator.

The monthly frequency of tropical cyclones in geographical squares has been shown in Figure 7. To complete the story, the annual variation of tropical cyclones needs to be shown. In Figure 9 this is done, and in Figure 10 the same thing is broken down to a five-year running average, covering a period of seventy-eight years. In Figure 10, a maximum at the start of the period will be noted, followed by another which started around 1930. Since this latter increase, the frequency has remained at a higher level, which accounts for the oft-repeated question, "Are hurricanes becoming more frequent?" Figure 10 shows that this has been true for the past few decades, and it fits with the fact that, over the same period, there has been a warming of the earth of some 2° F. and accompanying warmer ocean temperatures. This accounts for the sharks which now invade northern California beaches and for the codfish which are now found closer to Greenland than to the Grand Banks—all part of a general move to the north by sea and land animals.

When the climate is warmer than normal, the Bermuda-Azores anticyclone is north of its normal position and stronger than usual. This causes the southerly winds on the western end of the anticyclone to be of above-normal strength, thereby pushing hurricanes inland more frequently all the way from the Carolinas to New England. Old-time residents of these areas can vouch for the fact that there have been many more hurricane visitors than usual in the past thirty years.

There appears to be a slowing down of the warming trend, actually a change toward a cooling cycle. Dr. Hurd C. Willett (Massachusetts Institute of Technology), a leader in studies of long-range weather trends, predicts that the warm cycle will end and that in 1965 hurricanes will return to their normal tracks.

Movement

Every meteorology text in the past half century includes an illustration similar to Figure 11, showing principal hurricane tracks. About all this indicates is that hurricanes move along the southern periphery of the Bermuda-Azores High, turn northward part of the time, but head into the Gulf of Mexico when the high is pushing strongly over the eastern United States. These principal tracks are just averages of all tracks and are of no particular value in dealing with the erratic performers. It would be fine for everyone if the tracks for the past hundred years worked out statistically so as to provide ten to twelve standard tracks which hurricanes would follow

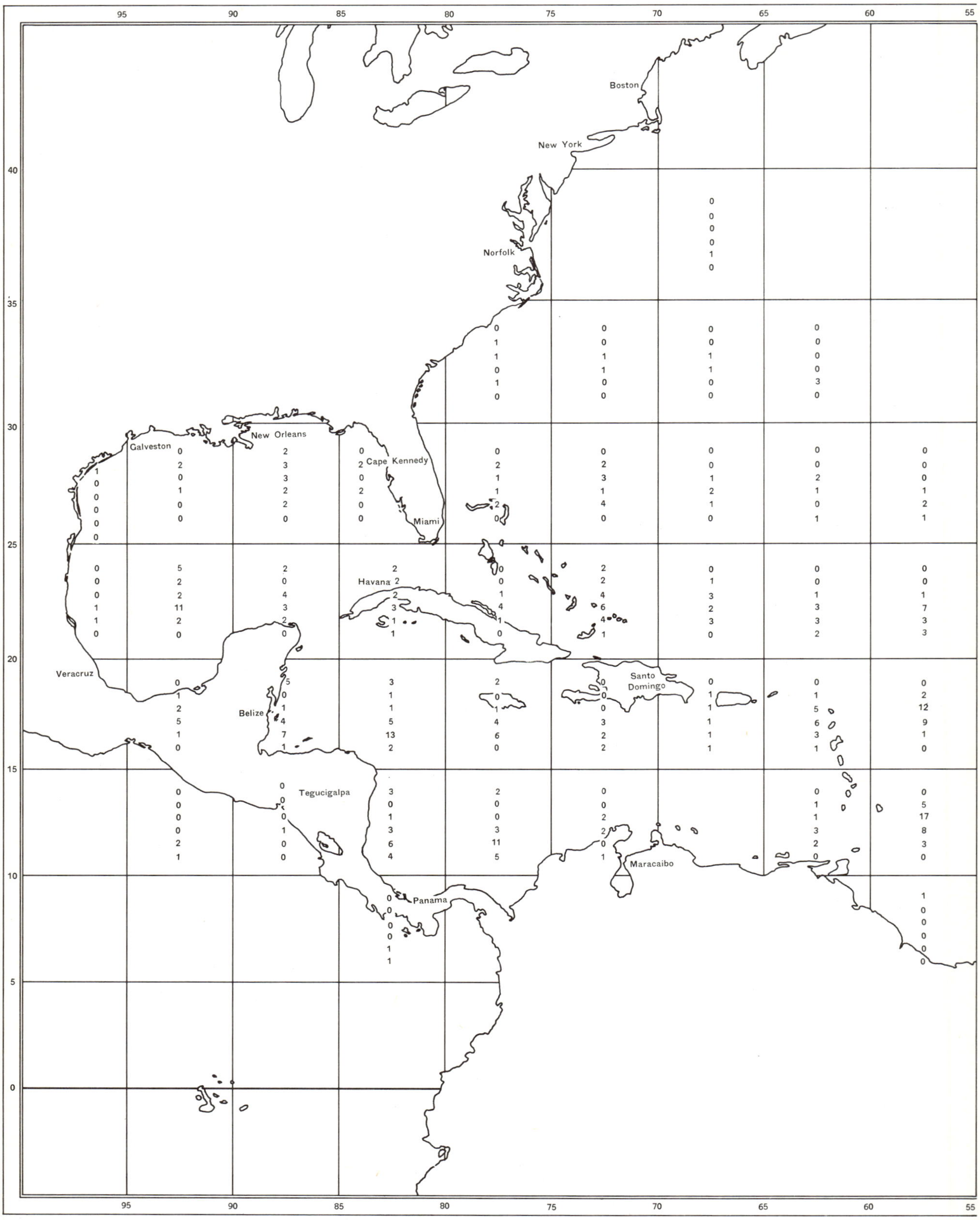

Fig. 7 Frequency of tropical storm origin by months within 5° areas, 1899-1958. (After Haggard.) Reading from the top, in each 5° area, the figures represent the number of tropical storms formed therein by months—June, July, August, September, October, and November-May.

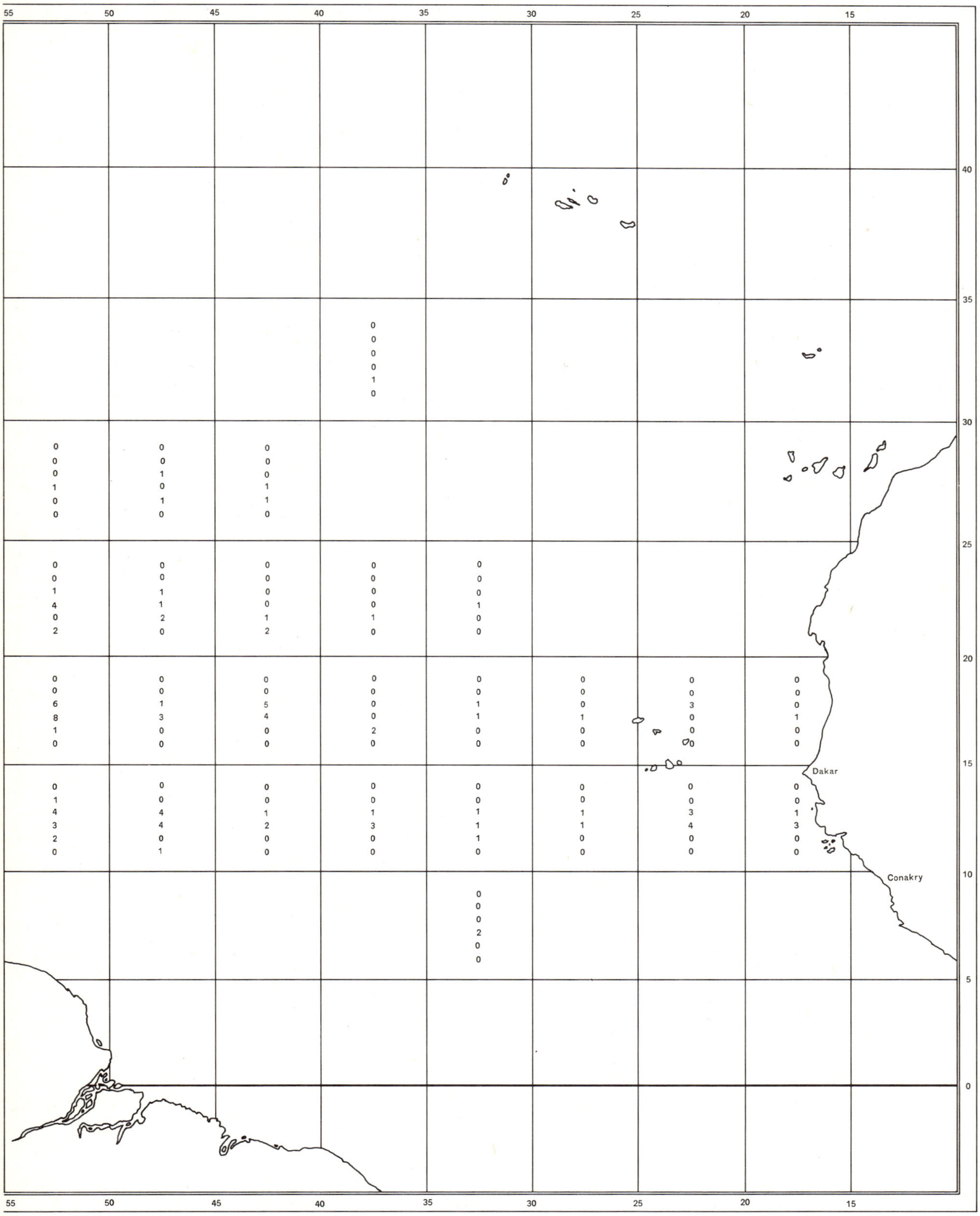
55
50
45
40
35
30
25
20
15
40
35
30
25
20
15
10
5
0
0 0 0 0 1 0
0 0 0 1 0 0
0 0 1 0 1 0
0 0 0 1 1 0
0 0 1 4 0 2
0 0 1 1 2 0
0 0 0 0 1 2
0 0 0 0 1 0
0 0 0 1 0 0
0 0 6 8 1 0
0 0 1 3 0 0
0 0 5 4 0 0
0 0 0 0 2 0
0 0 1 1 0 0
0 0 0 1 0 0
0 0 3 0 0 0
0 0 0 1 0 0
0 1 4 3 2 0
0 0 4 4 0 1
0 0 1 2 0 0
0 0 1 3 0 0
0 0 1 1 1 0
0 0 1 1 0 0
0 0 3 4 0 0
0 0 1 3 0 0
0 0 0 2 0 0
Dakar
Conakry

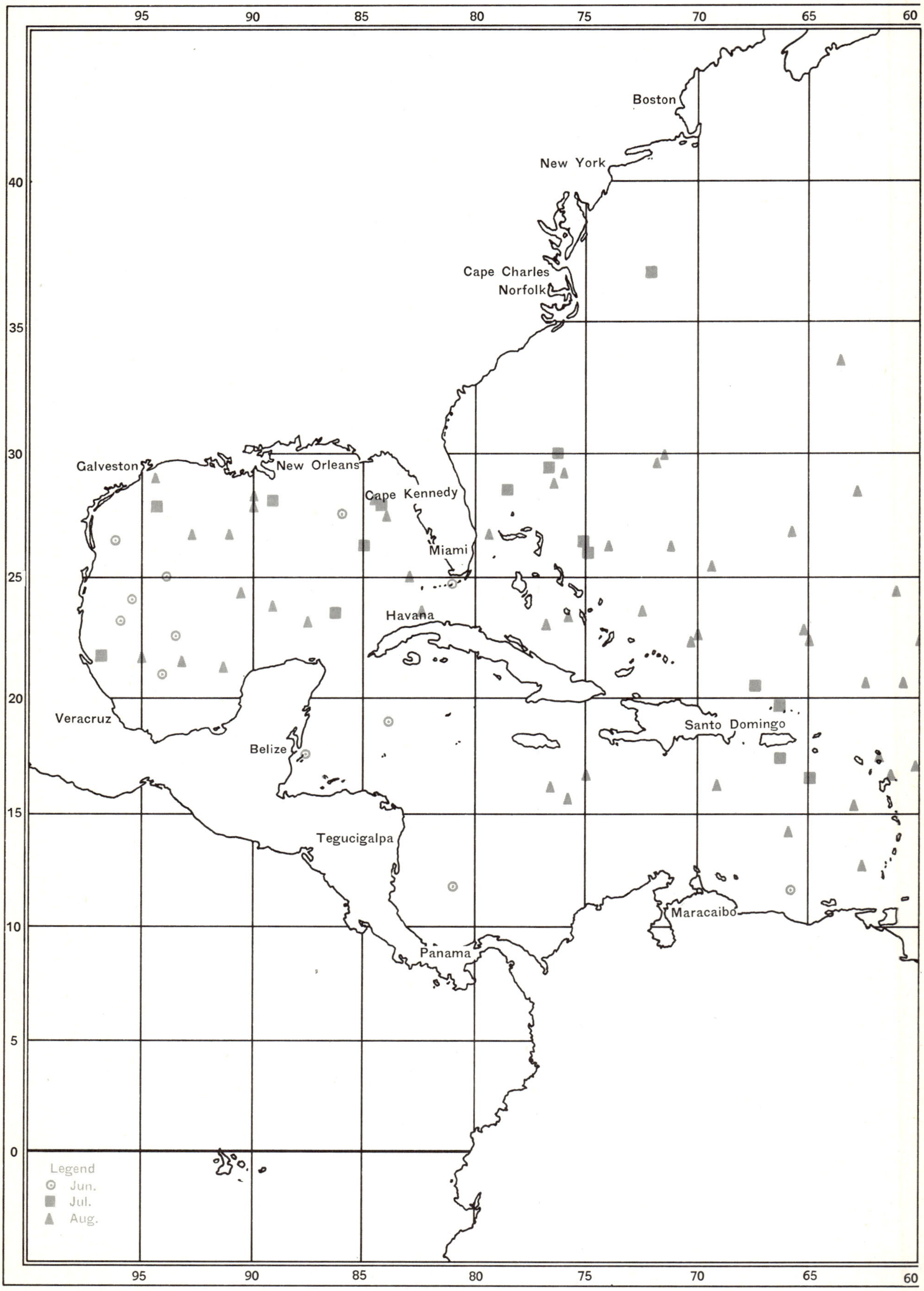

Fig. 8A Locations where tropical cyclones reached hurricane intensity during the months of June, July, and August, 1899-1957. (After Haggard.)

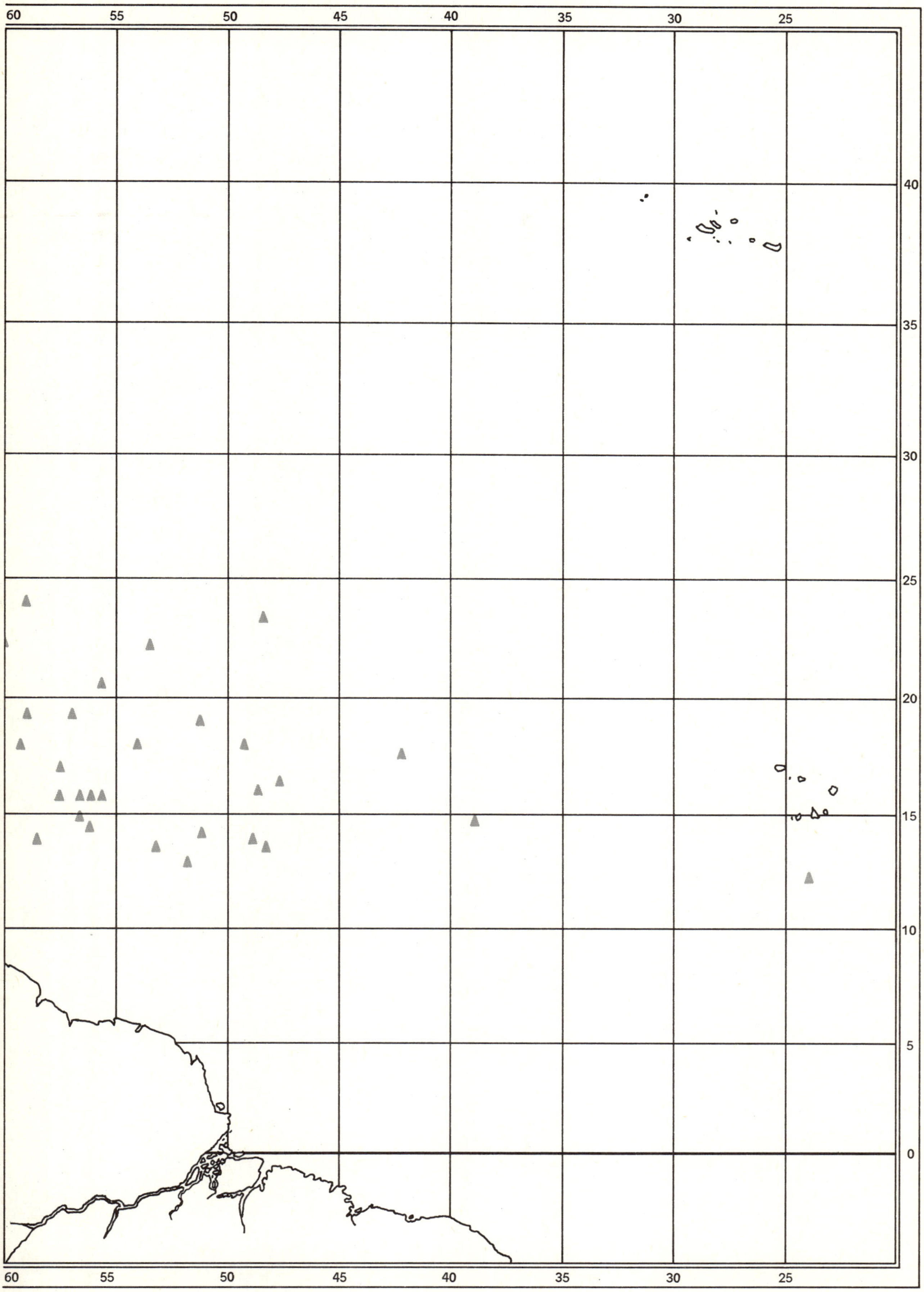
60
55
50
45
40
35
30
25
40
35
30
25
20
15
10
5
0
60
55
50
45
40
35
30
25

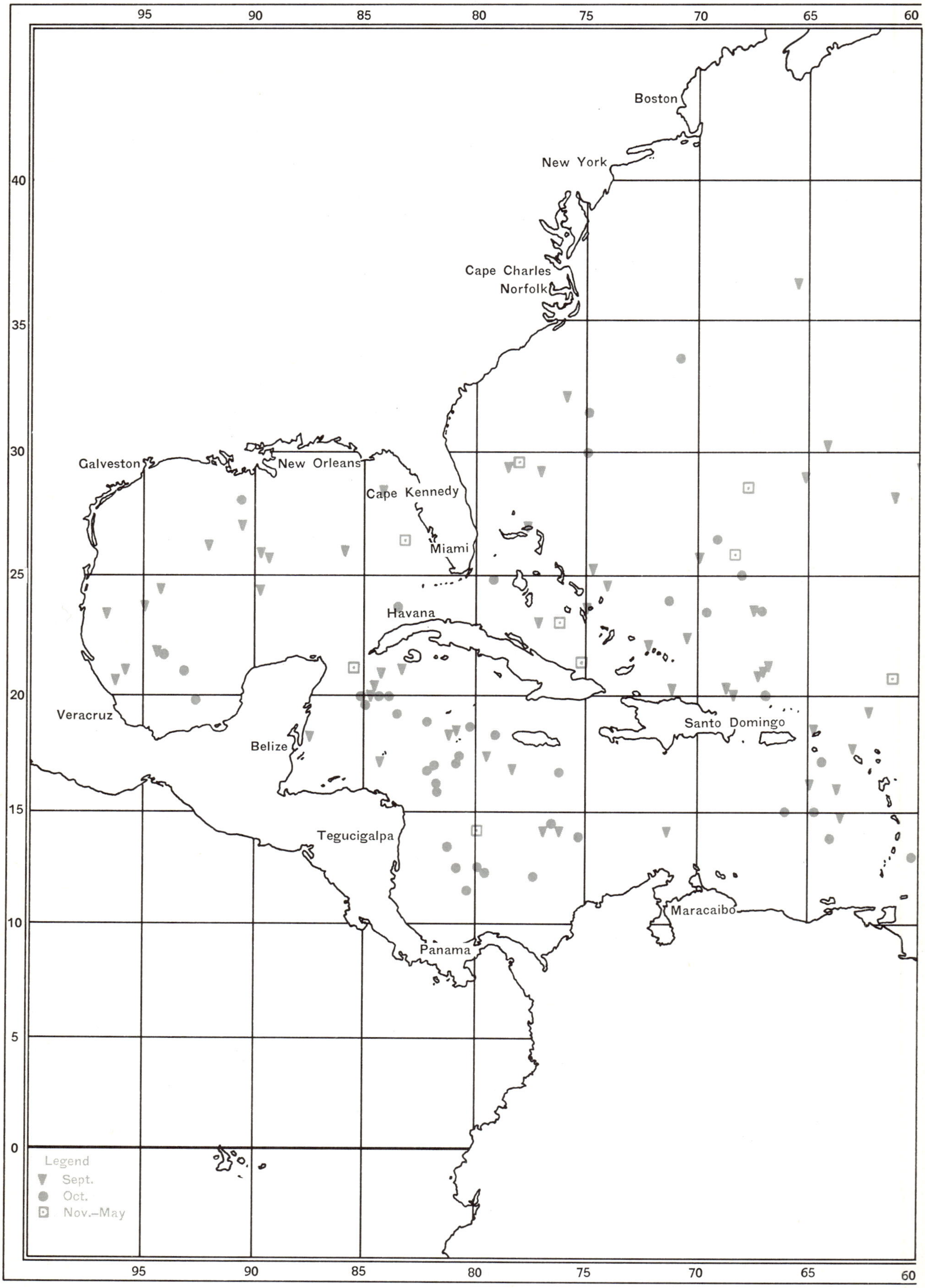

Fig. 8B Locations where tropical cyclones reached hurricane intensity during the months of September, October, and November, 1899-1957. (After Haggard.)

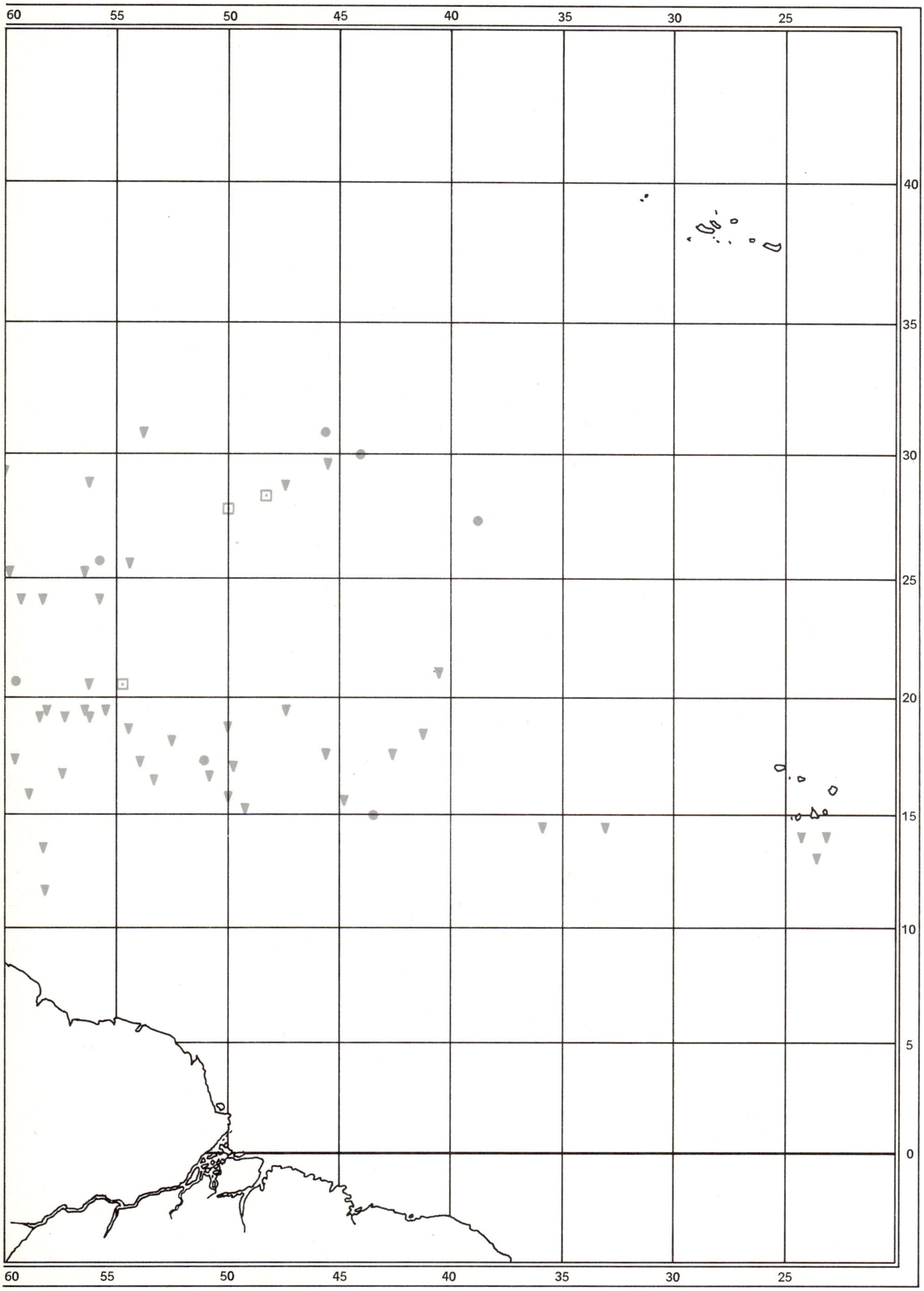
60
55
50
45
40
35
30
25
40
35
30
25
20
15
10
5
0

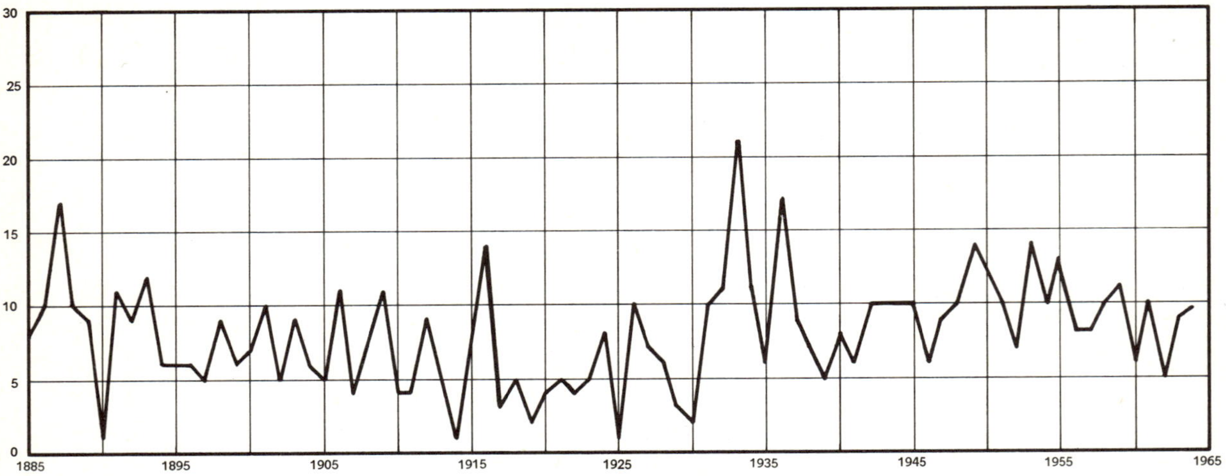

Fig. 9 Annual variation in the number of tropical cyclones.

Fig. 10 Five-year running average of tropical cyclone frequencies.

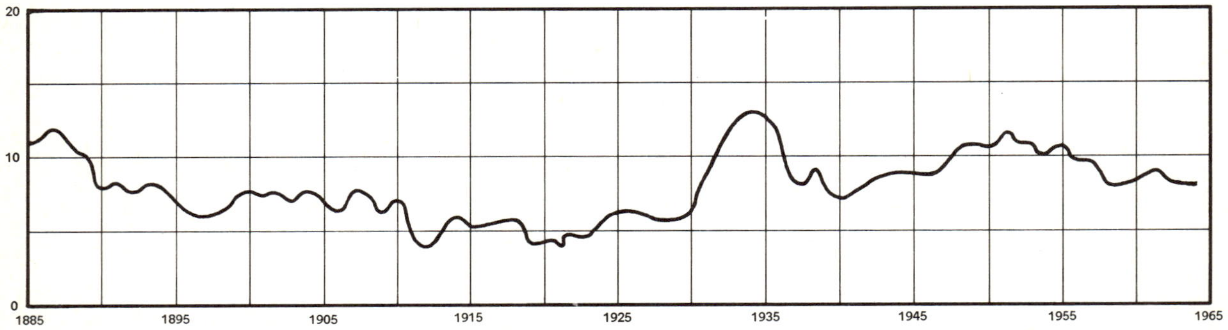

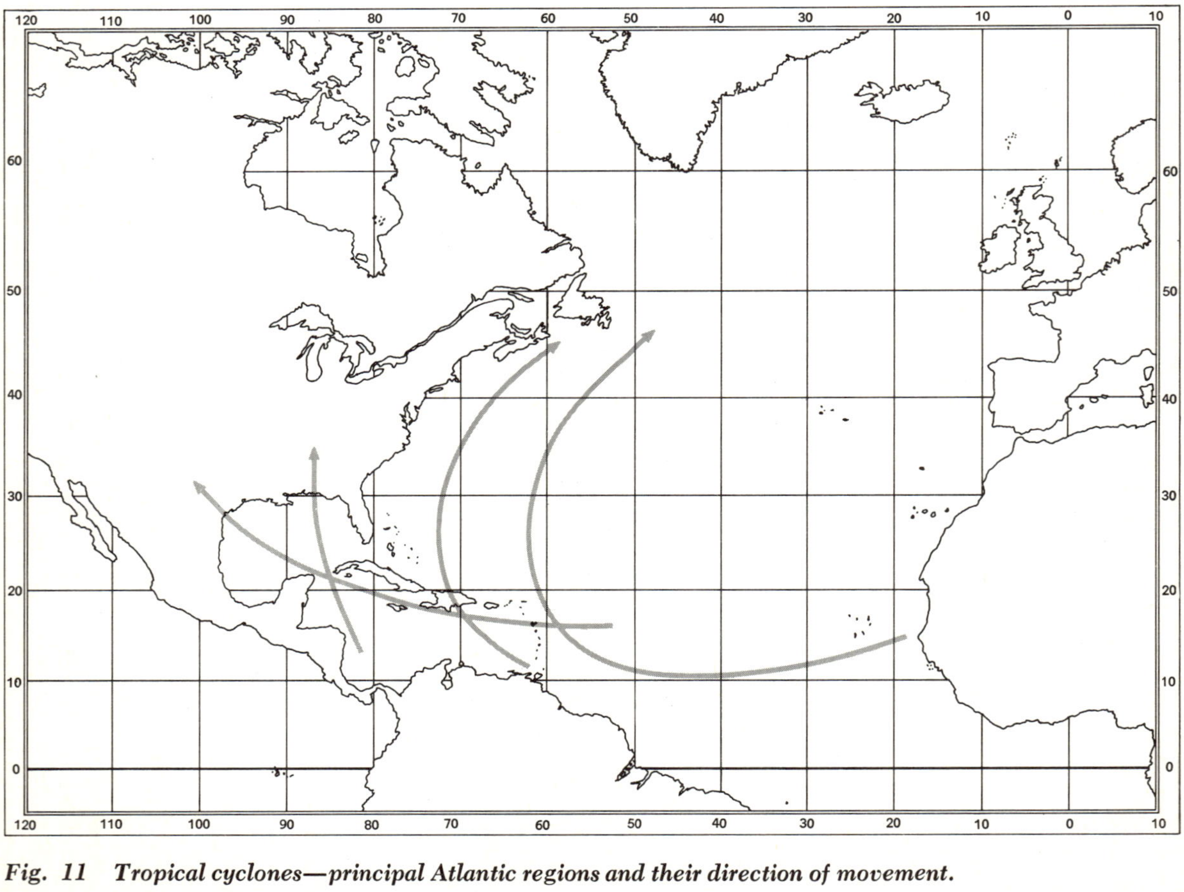

Fig. 11 Tropical cyclones—principal Atlantic regions and their direction of movement.

with a high degree of probability. Warnings could then be sent out saying, "This hurricane is on Track #8, moving at 14 knots." Ships and shore stations could break out the book and see how this might affect them.

It does not work out this way. Look, for instance, at the tracks for September, Figure 12. True, there is a general similarity of movement in the early stages of the storms, along the ESE-WNW axis. But the turn to the north occurs with no longitudinal regularity, and among the tracks you see the benders, loopers, double benders, and wobblers. Not apparent in the figure are the storms with a change of pace, with their exasperating sudden accelerations (as much as 1500 per cent in 24 hours) and their sudden decelerations; nor do we see the centers which stop suddenly and sometimes hold position within a 50-mile circle for as much as three days. Figure 14, page 90 in Part II—Typhoons, illustrates twelve basic typhoon tracks. These are true for hurricanes as well. Captain Kotsch has given the tracks names to match their behavior. These tracks are well worth a look.

At this point, one might ask, "Is there anything in all this track confusion which I can use operationally, or shall we get on with something worthwhile?" The answer, "There is!"

First, there is a definite hurricane season which extends from June to mid-November. It is true that there have been a few storms which formed in December and January, but these cases are too infrequent to worry about.

Second, within the annual June to November season, there is a month-by-month movement pattern. Starting in June, storms develop in the western Caribbean or in the Gulf of Mexico and move northwestward, with the Texas coastline the most likely target at this time of year. During August and September, hurricanes develop over the eastern North Atlantic south of the Cape Verde Islands. The majority of these storms move in a westerly direction across the Atlantic; some reach the United States mainland before they recurve to the north and northeast. In the latter part of September, and in October and November, hurricanes again originate in the western Caribbean Sea. In September, they follow much the same paths as those of June, but in October and November they are more likely to turn northward and northeastward in lower latitudes, sweeping out over Florida or the Greater Antilles.

For operational planning, the picture can be narrowed even more. Hurricanes forming in the same general area at the same time of year have preferred tracks, although there are exceptions. Figure 13 shows the percentage-frequency distribution of the direction of motion of tropical cyclones by 5° squares for the several months of the hurricane season. From this figure, the most probable direction can be determined, as well as probability of displacement in other directions. In using these charts, Dr. José A. Colón (U. S. Weather Bureau), the originator, suggests these thumb rules:

1. Hurricane displacement probabilities, because of the large number of cases, are quite reliable in August, September, and October.

2. In August, no storm of record south of 25° N and east of Florida has moved east of north or south of west.

3. In some areas, particularly the Gulf of Mexico, there is no reliable most probable direction of movement. This is where you hope the hurricane forecasters are correct.

4. The charts are of most value when the storm is very young and at the highest and lowest latitudes. This is fine, except that the area between 20° to 30° N, where the charts are least reliable, is just where they threaten the United States coastline and the many ships to be found in the coastal shipping lanes.

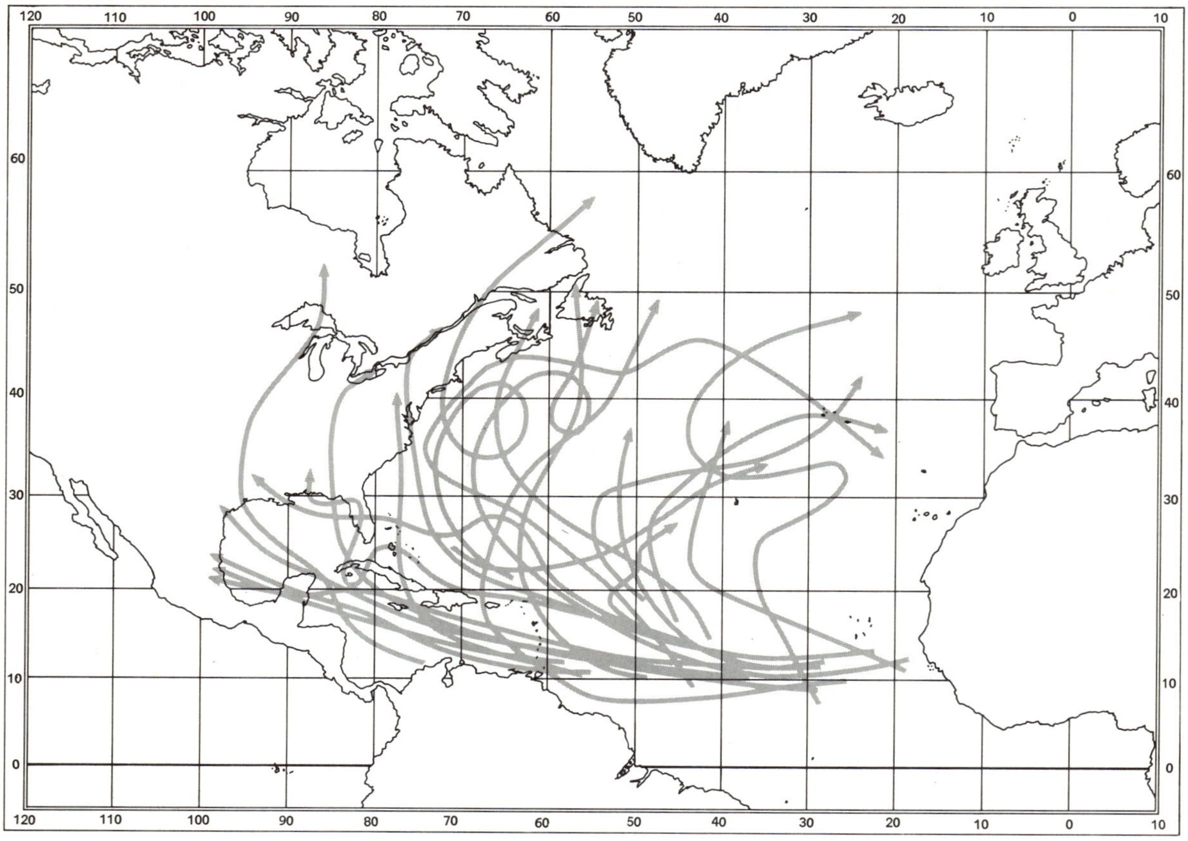

Fig. 12 Selected hurricane tracks for September.

Examining a few movement quirks will make it evident that one should not place too much confidence in the climatological charts:

1. The mean charts tend to smooth out action by the speeders. The mean speed of hurricane movement south of 30° N varies by areas between 12 to 16 knots, except for November when the 20-knot mean speed line dips southward a few degrees. When the Bermuda-Azores High is particularly strong, there will be some hurricanes which will move at 20 to 25 knots south of 30°N. Based on continuity alone, these are easily forecast. But north of 30° N, trouble starts, and the speed range increases from 0 to 60 knots. Acceleration can be pronounced, and there is a strong tendency among forecasters to set the speed too low—with disastrous results in some cases. New England has been the most frequent victim in this respect. Hurricanes that were loafing along off the Virginia Capes have suddenly spurted and roared up the coast, sometimes at speeds as high as 70 knots.

This lesson was well learned on one occasion with a storm moving ENE at a point about 150 miles east of Cape Cod. The storm speed of movement had been moved up to 42 knots, although this seemed much too fast. A destroyer skipper got one warning a few hours after issuance and sent a pointed dispatch which started, "You may think the hurricane is at 42° N and 66° W. At your warning time, my ship was at 42.3° N and 63.7° W, and. . . ." He went on to describe his 75-knot winds and the passage of the eye.

2. Hurricanes favor movement over the warmest waters. If there are doubts about which way to head a ship, charts showing the sea surface temperature distribution should be checked and the warm water areas should be avoided. The Gulf Stream on many occasions has served as an inviting path.

3. In their early stages, hurricanes may not move in a straight line. Instead, they wobble back and forth in a sinusoidal path. The amplitude of the hurricane swings is nothing stupendous. One, for instance, had a swing of 15 miles from side to side, a 130-mile wave length, and crossed the mean axis of movement direction every five hours. When following a storm on radar, either at sea or ashore, one is likely to see a storm of this nature heading in your direction for a few hours and then turning away. This sort of thing can be hard on one's nerves, even when you know it can happen.

4. Should warnings on two storms be received, a good look should be taken to determine how close together they are. These warnings may indicate the likelihood of the Fujiwhara effect, so named for the well-known Japanese meteorologist who studied and described the influence which two tropical cyclones in close proximity (1000 miles or less) tend to have on each other's motion. To get an idea of the general effect, visualize the two storms as the ends of a giant dumbbell, with an all-powerful hand grasping the joining bar and turning the dumbbell counterclockwise. The storms tend to rotate around each other. The mid-point or center of the joining bar also moves, as though the hand were lifting slowly as it turned. The relative motion is further complicated by the fact that the storms have mutual attraction; they get closer together with time, and the closer they come together, the faster they tend to rotate about each other.

The condition of having two hurricanes in existence at the same time, and close together, is rare. It occurs much more frequently with typhoons in the Pacific, and Captain Kotsch mentions on page 135 a case where a Navy ship sailed safely through the "bar of the dumbbell." Double storms do occur in the Atlantic, however. On one occasion there were four storms showing on the weather map.

Fig. 13 This drawing and the following five depict the percentage-frequency distribution of motion of tropical cyclones by 5° squares for the months of June through November. A 22.5° sector is included with each percentage line. (After Colón.)

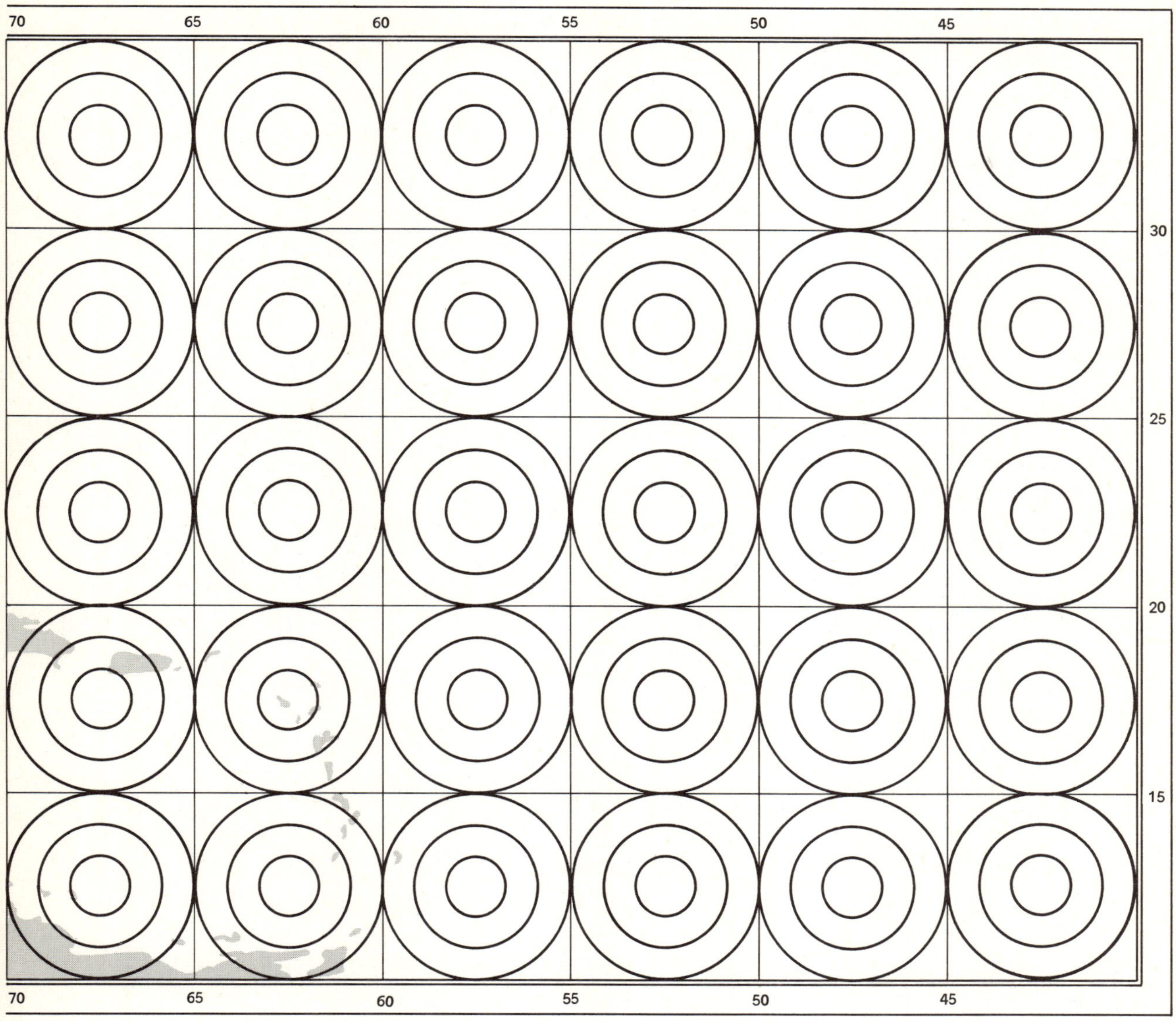
70
65
60
55
50
45
30
25
20
15

JULY

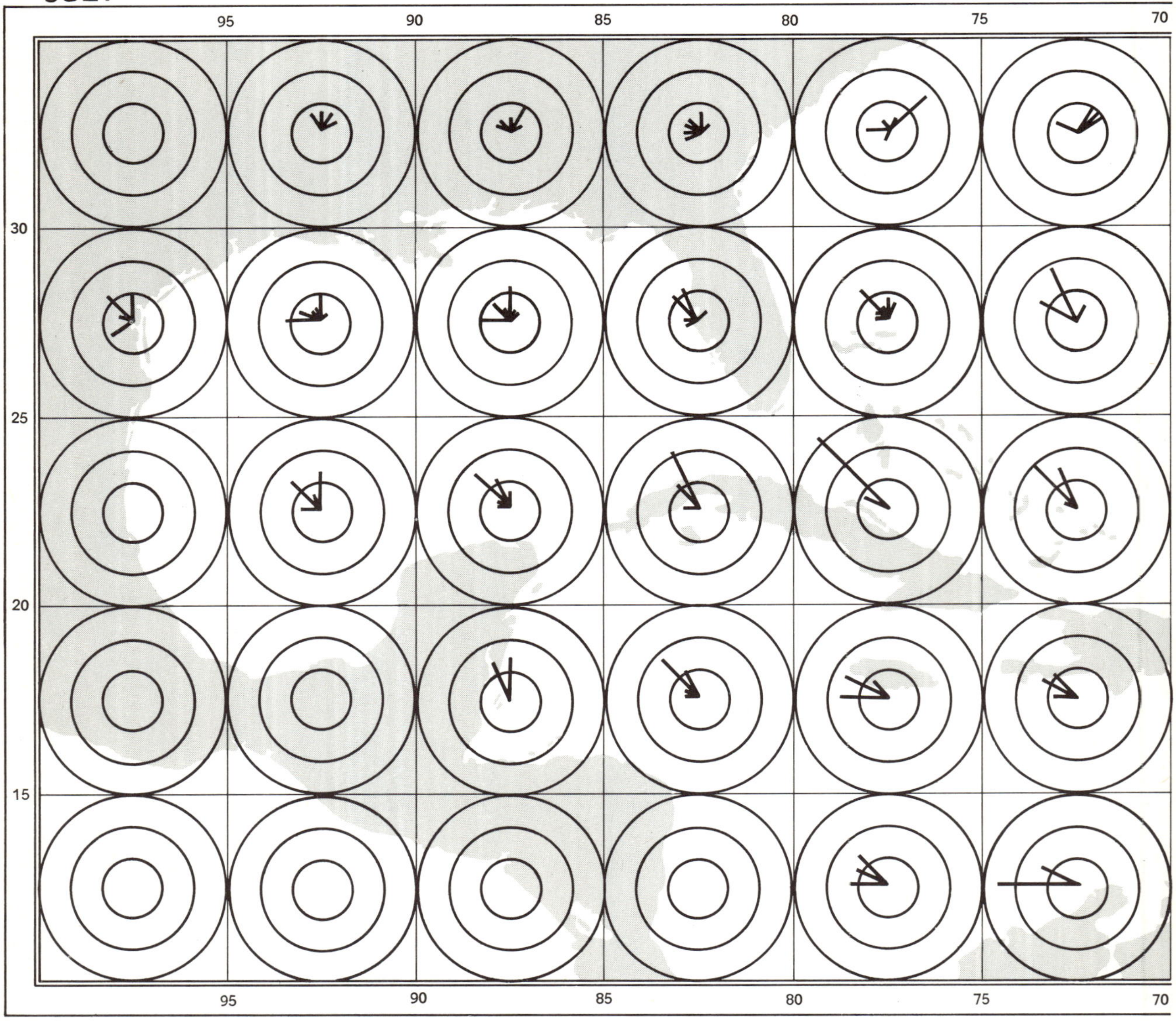

LEGEND
Scale Percent
25%
50%
75%
Direction of lines indicates direction of motion of tropical cyclones.

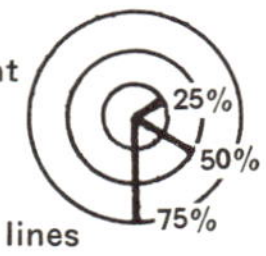

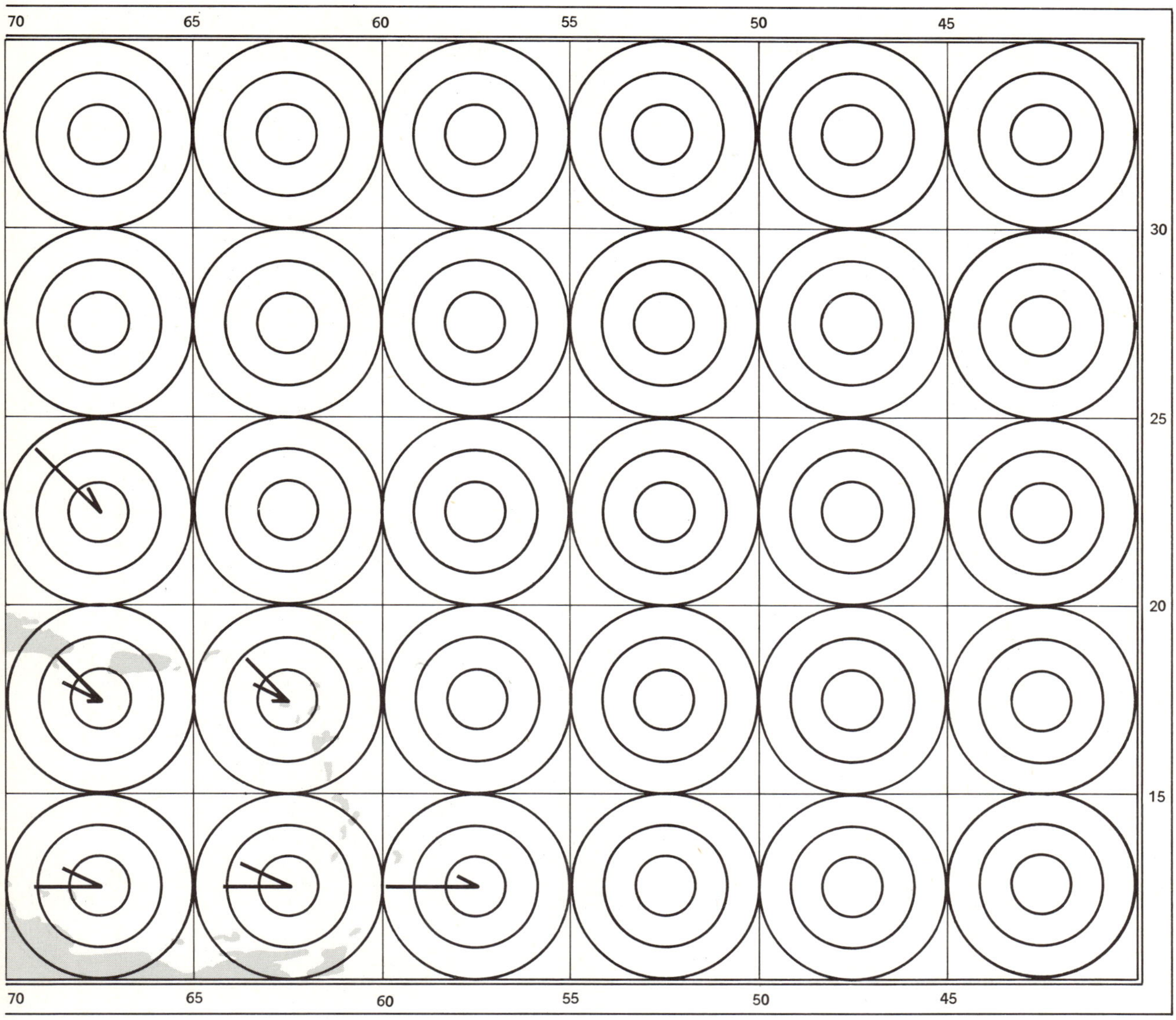
70
65
60
55
50
45
30
25
20
15
70
65
60
55
50
45

AUGUST

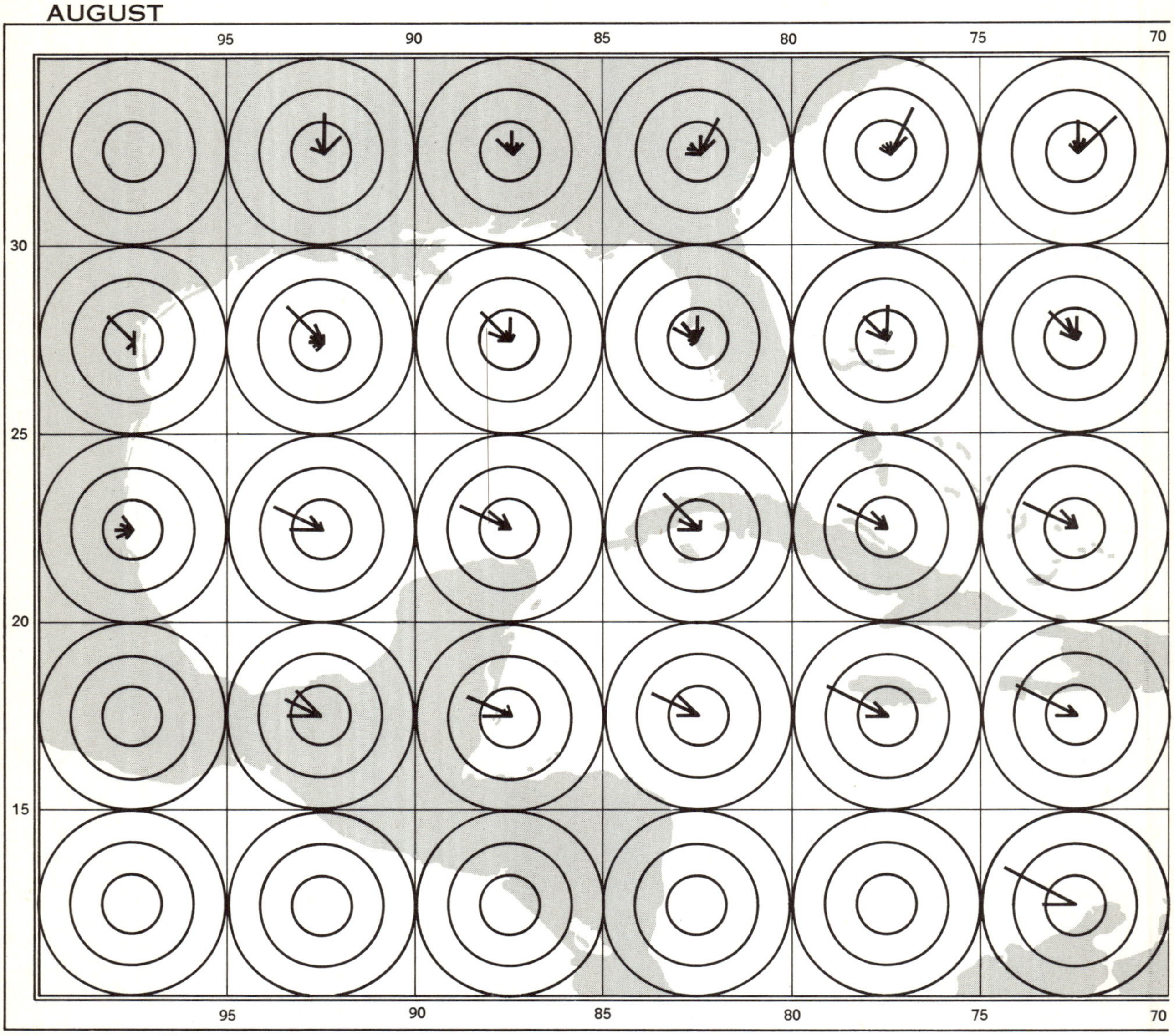

LEGEND

Scale Percent

25%

50%

75%

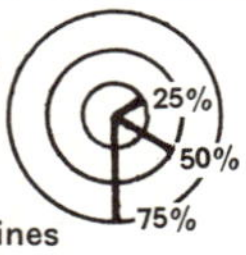

Direction of lines indicates direction of motion of tropical cyclones.

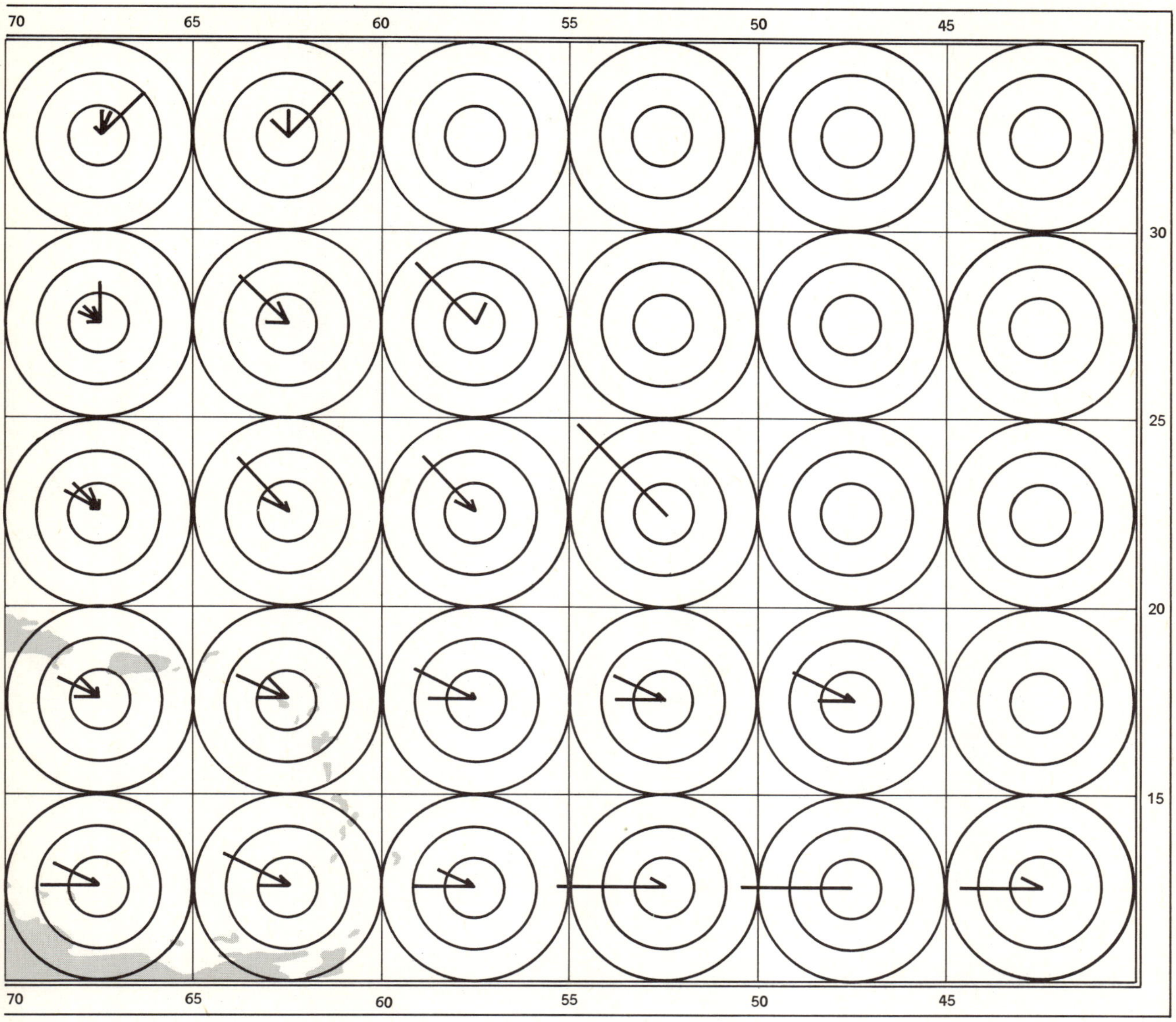
70
65
60
55
50
45
30
25
20
15
70
65
60
55
50
45

SEPTEMBER

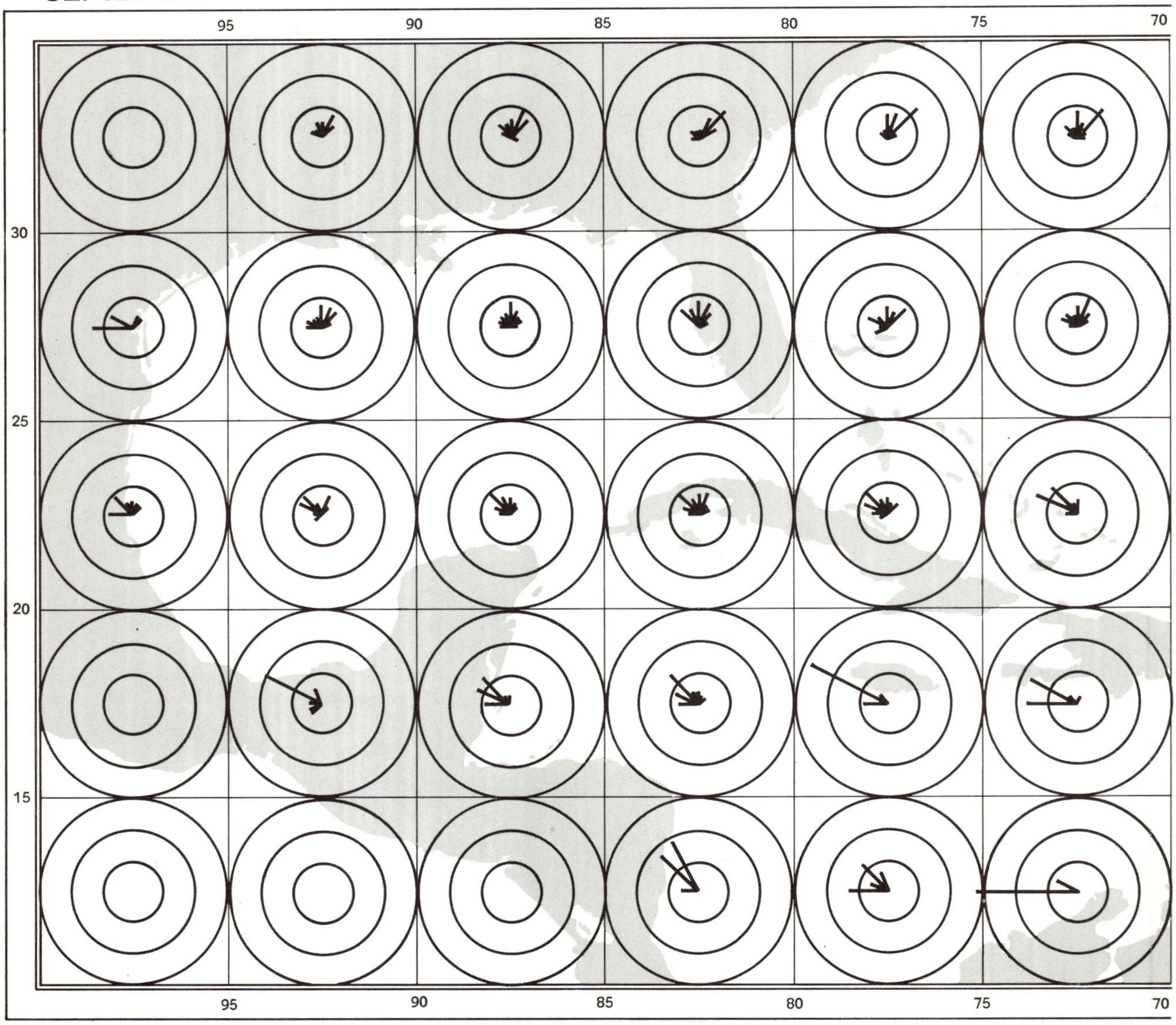
95
90
85
80
75
70
30
25
20
15
95
90
85
80
75
70

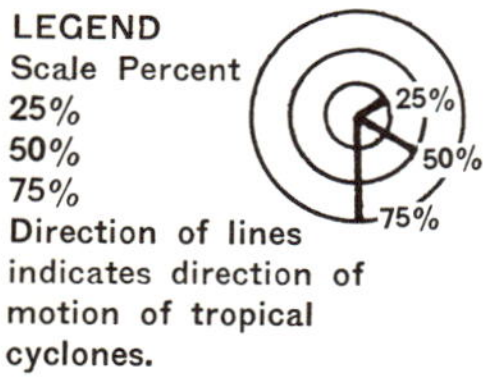
LEGEND
Scale Percent
25%
50%
75%
25%
50%
75%
Direction of lines indicates direction of motion of tropical cyclones.

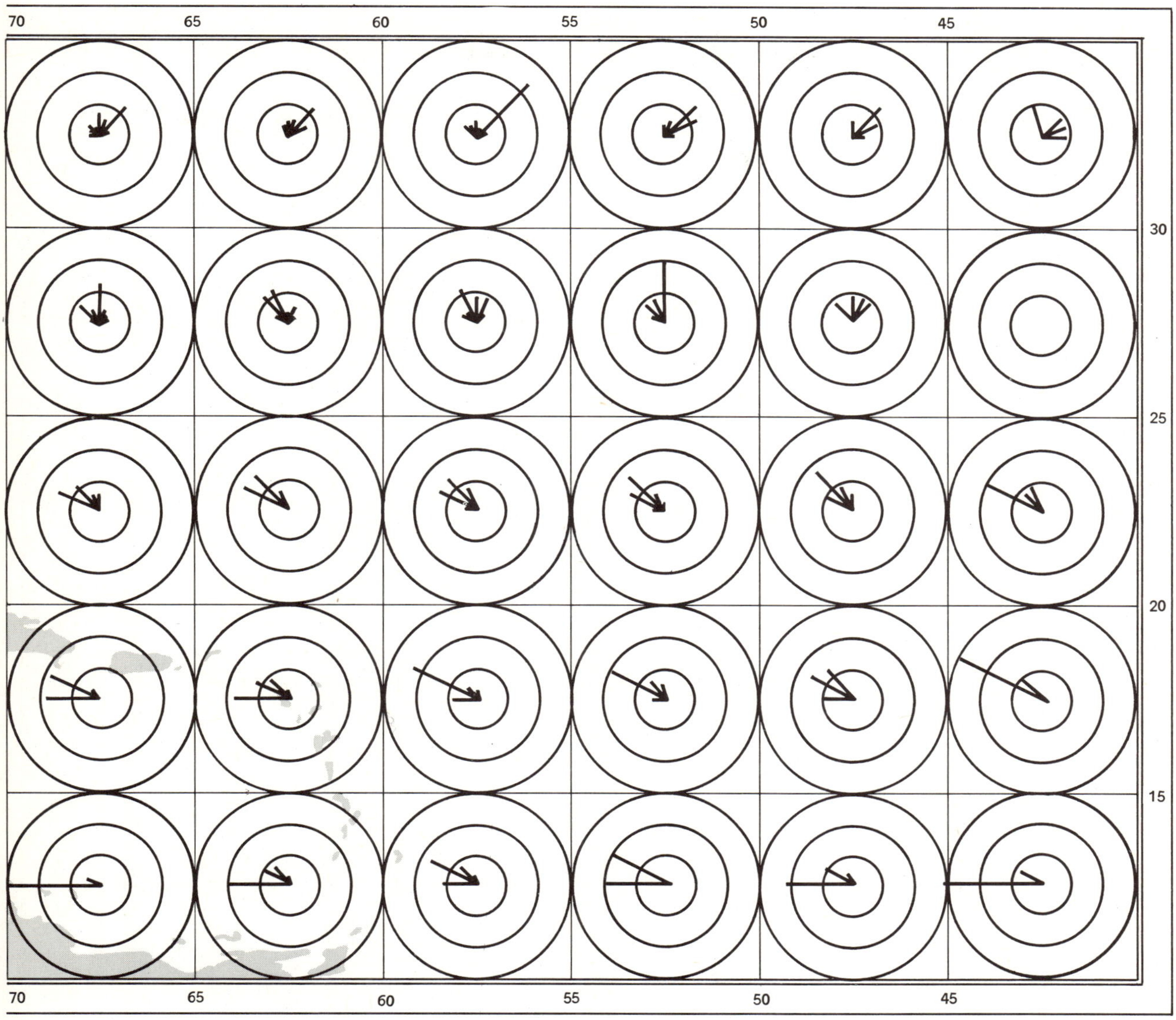
70
65
60
55
50
45
30
25
20
15
70
65
60
55
50
45

OCTOBER

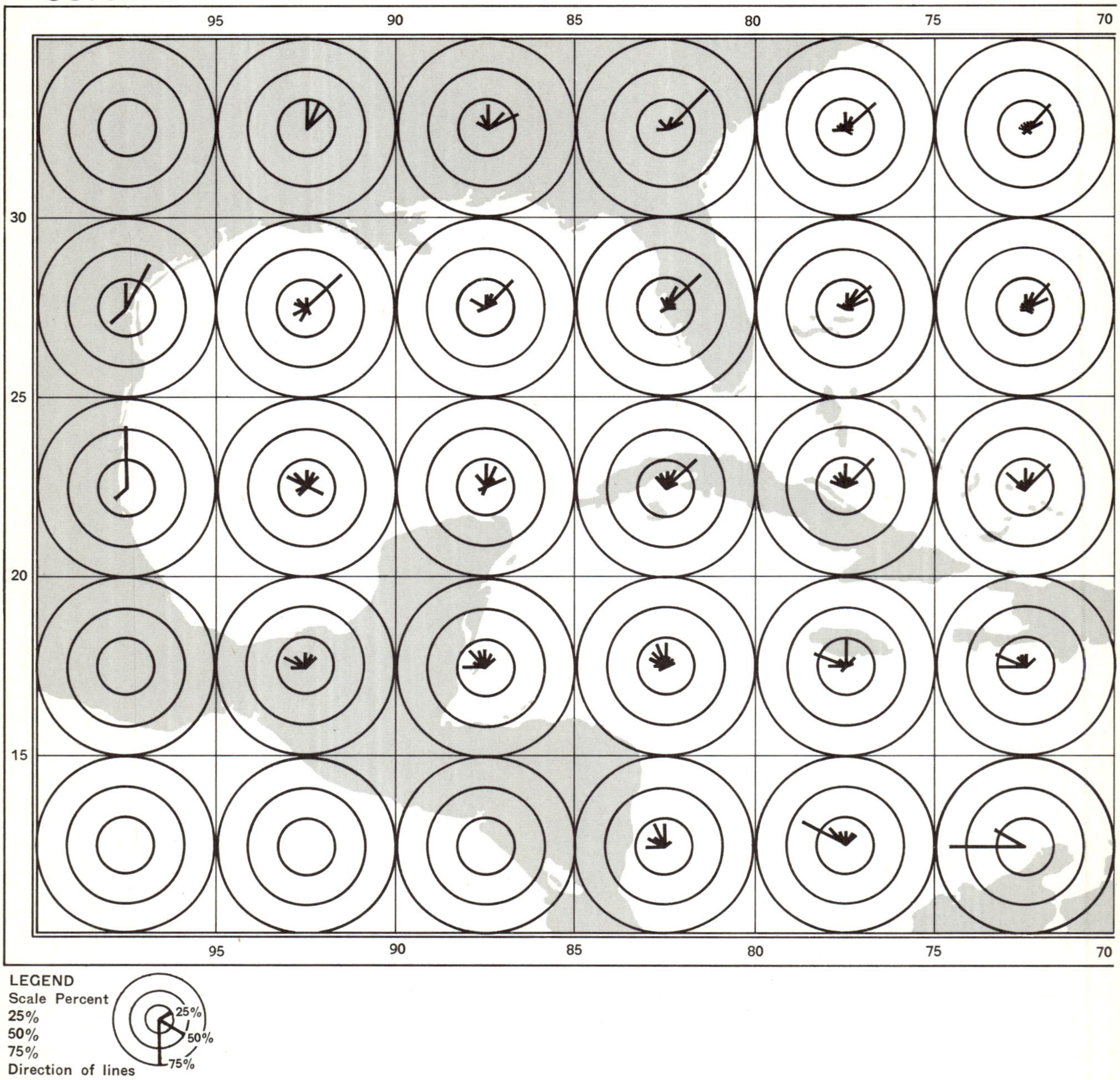

Direction of lines indicates direction of motion of tropical cyclones.

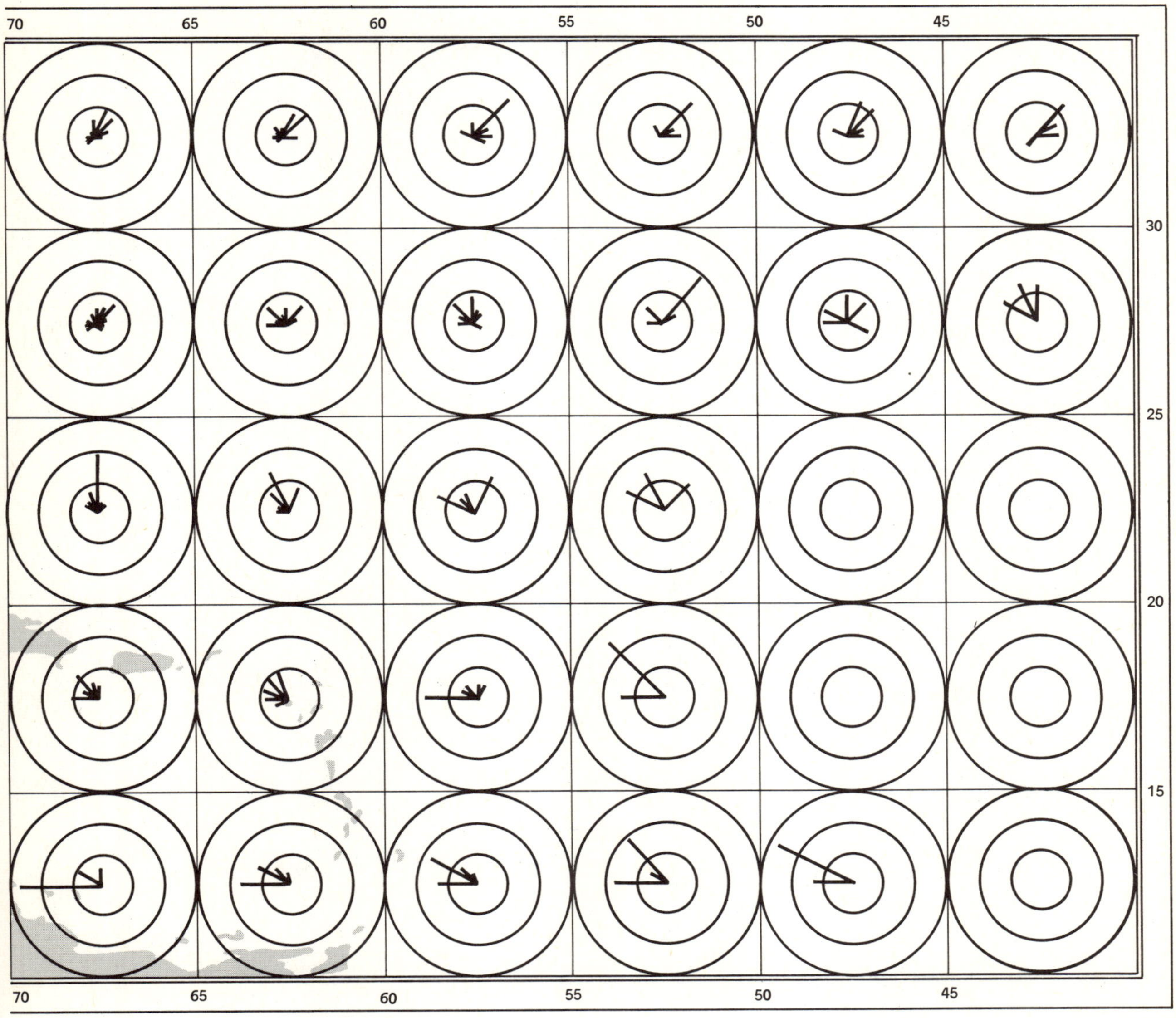
70
65
60
55
50
45
30
25
20
15

NOVEMBER

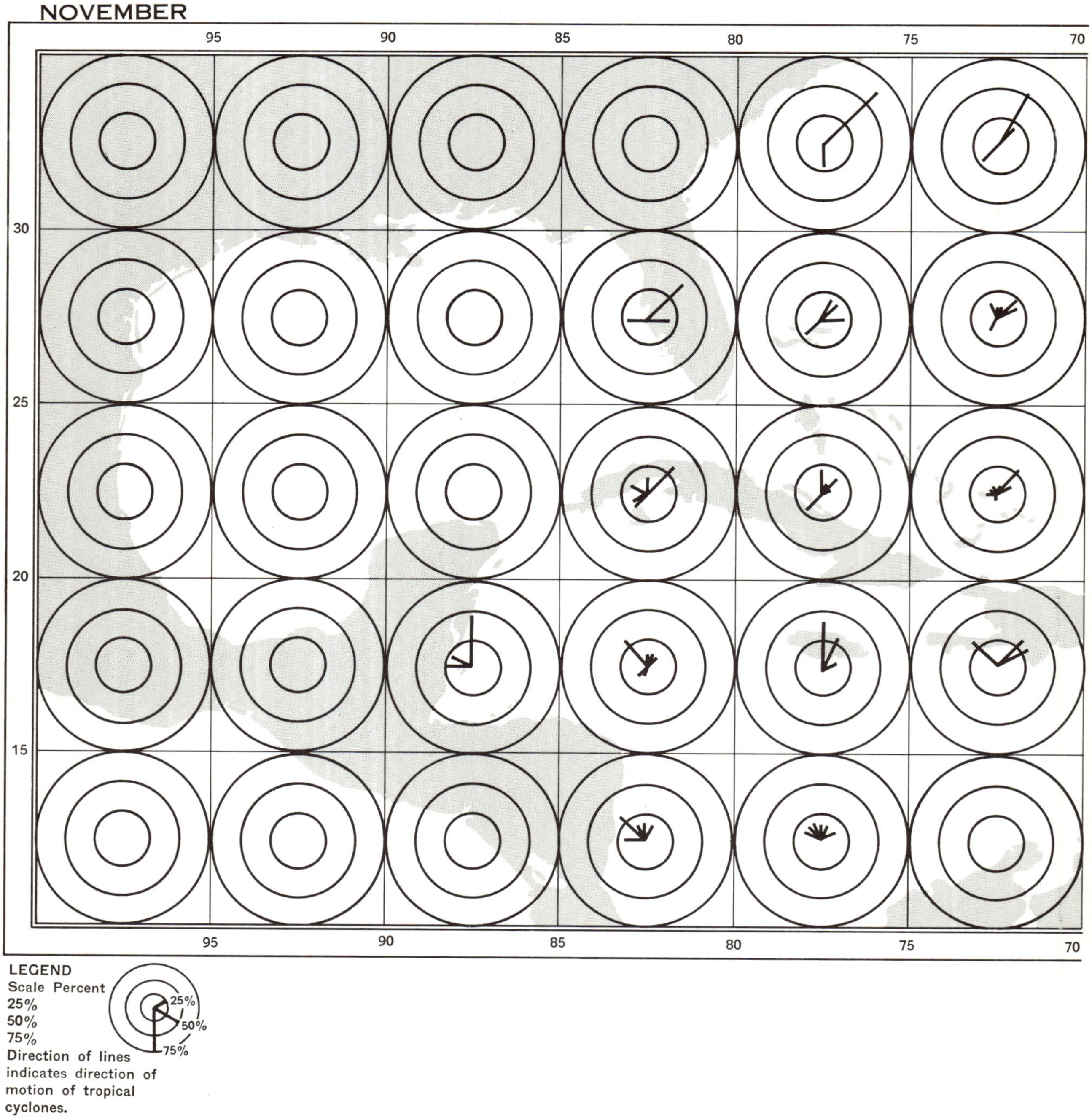
95
90
85
80
75
70
30
25
20
15
LEGEND
Scale Percent
25%
50%
75%
25%
50%
75%
Direction of lines indicates direction of motion of tropical cyclones.

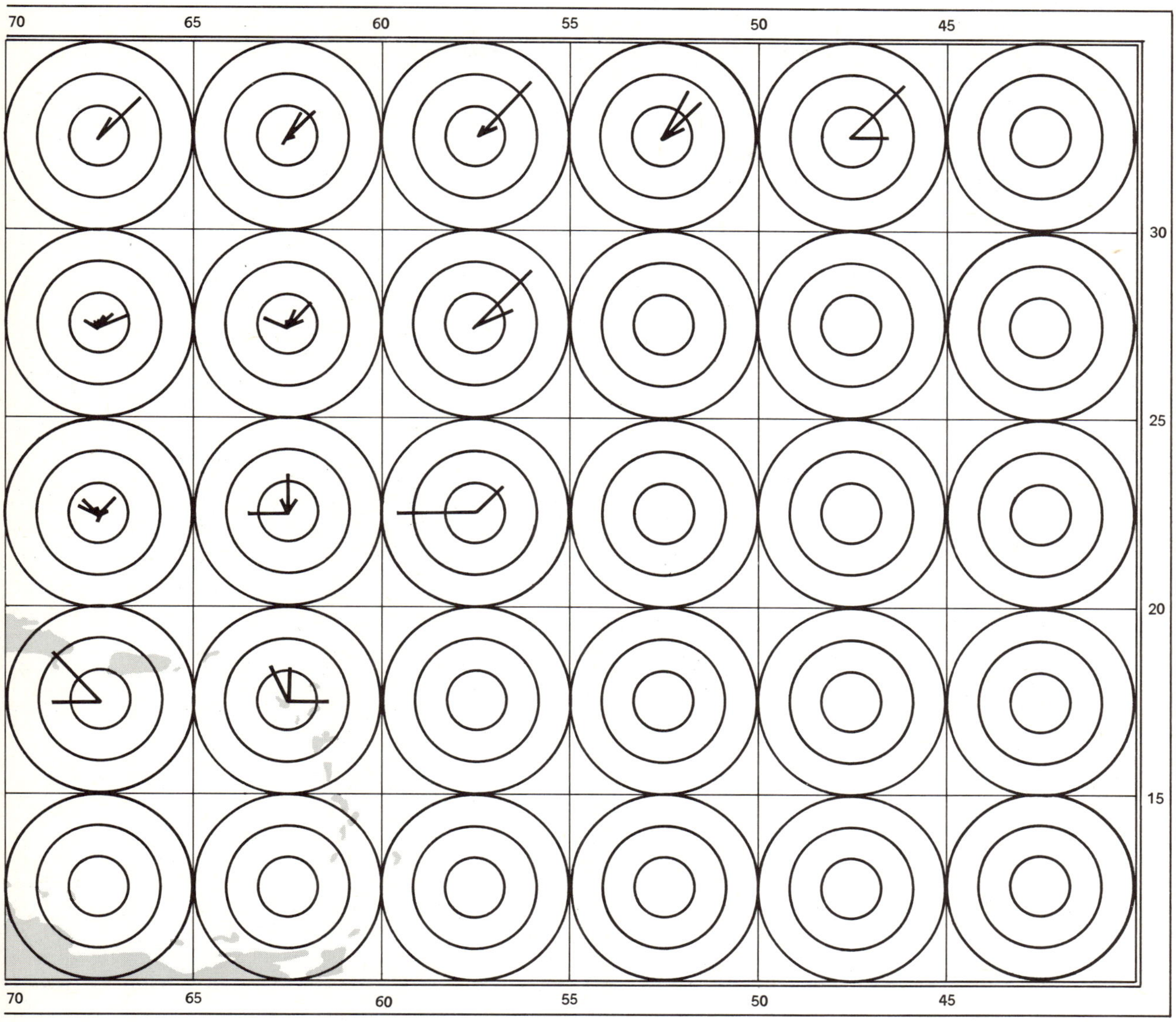
70
65
60
55
50
45
30
25
20
15

Section 3 : Characteristics of the Hurricane

Hurricanes have been described in literature and sea stories for centuries in the strongest, gaudiest of language, but description, no matter how high-blown the superlatives, has no operational value unless the characteristics can be described in numerical terms.

Certain basic features of hurricane structure and behavior should be kept in mind:

Hurricane winds blow in a counterclockwise direction around the central eye in the Northern Hemisphere.

Each hurricane seems to have individual characteristics, whether of intensity, movement, or size. Hurricane climatology supports the often repeated statement that no two hurricanes have ever been exactly alike.

Hurricanes always originate in the tropics and over water.

The energy necessary to keep the hurricane circulation alive comes from the warm water surrounding the eye. That it does not come from the area near the eye was vividly demonstrated in Hurricane Flora (1963), whose eye was over Cuba for 24 to 36 hours with no marked diminution in Flora's intensity.

Size

Hurricane size is expressed in several ways—strength of maximum winds, diameter of hurricane-force wind area, diameter of gale-force (34 knots minimum) wind area, and finally, the over-all size of the cyclonic circulation as measured by the diameter of the outer closed isobar (line of equal pressure) on the weather map. Description in official warnings includes the first three definitions.

The highest winds of the hurricane have never been accurately measured, but from reasonable estimates we know that most hurricanes reach 90-knot intensity sometime during their lives, and that some probably go as high as 170 knots (200 mph). The 90-knot intensity would represent the mean, so a 130-knot hurricane is *big*.

The area of hurricane force winds (64 knots or 75 mph) averages a little over 100 miles in diameter; in larger storms, the area may be as much as 200 miles, the distance between New York and Washington.

Gale force winds may extend over an area 600 miles in diameter; the average would be about 400 miles.

Unfortunately, correlation is poor between maximum winds near the hurricane center and the diameter of the strong wind area. It can not be said with any assurance that a hurricane with 40-knot winds 150 miles away from the center will have peak winds of 125 knots near the center, any more than it can be said that a 125-knot center means winds of 50 knots at any specified distance from the center. For example, the Florida Keys hurricane of 1935, the most intense on record, had a path of destruction only about 40 miles wide; some storms have had even smaller areas of very high winds. In general, storms of small diameter have winds as intense near the center as some of the giants whose gale force winds extend outward for hundreds of miles.

Wind

The winds of the hurricane are its most spectacular feature, although the seaman might cast a vote for the heavy seas, and some others for the eye. The strongest winds have probably never been measured because the wind-measuring devices are not designed to stand much more than about 125 knots, after which they either stop functioning or "are gone with the wind." It may be, too, that the area of strongest winds in a hurricane never passes over a measuring instrument ashore or at sea. However, from reconnaissance reports, or from engineering calculations based on damage, we can get a reasonably accurate estimate of the highest winds in most storms.

Reconnaissance planes have often reported winds in the 130- to 150-knot (150 to 175 mph) range; actual measurements at land stations have been made as high as 150 knots; and in the storm that struck the Florida Keys in 1935, engineers estimated that the winds must have been in the 170- to 215-knot (200 to 250 mph) range to account for the damage they did.

One feature of hurricane winds which accounts for much of the wind damage is the gustiness. If the winds would start blowing slowly, gradually and steadily build up to peak force, and go back to normal in the same steady fashion, structures would be much better able to withstand the forces involved. But, from observation and calculations, it is known that momentary gusts exceed the steady winds by 30 to 50 per cent. A wind of 100 knots may have momentary gusts to 150 knots. So, not only is there this temporary additional wind force with which to deal, but also a force applied unsteadily in a process similar to that involved with rocking a car back and forth to get it rolling.

Figure 14 shows the wind circulation around a typical hurricane in the Northern Hemisphere and reveals two additional characteristic features of hurricanes. The winds do not blow straight along the isobars; instead, they angle toward the center between 20 to 30 degrees. This angling-in occurs all the way from the outer limits of the hurricane circulation up to the wall of the eye. The angle grows less and less as we approach the eye, and the winds grow stronger and stronger. Finally, at the eye wall, the winds blow right along the isobars, just as though they were moving around the outside of a round bottle. The inflow of 20 to 30 degrees accounts for the fact that birds, butterflies, and helpless ships gradually drift into the eye of the hurricane.

The second characteristic to be noted on Figure 14 is that the winds are stronger along the right-hand side of the direction of motion. This is readily understandable when you consider that the body of air which makes up the hurricane is moving at a certain speed in a certain direction. If this speed should be 20 knots, that amount is added to the wind speed in the side of the hurricane to the right of the direction of motion, and it is subtracted from the other side. If a small aircraft should be flying on the strong side of this hurricane, where the winds are 100 knots, 20 knots of this wind would be due to the motion of the storm. Were our intrepid flyer to shift across the eye—if he made it and if the storm were symmetrical—he would find that he would be in 60-knot winds.

Pressure

Any reader of tropical yarns knows that trouble lies ahead when the mate on watch says to his assistant, "Wake the Captain and tell him the bottom's dropping out of the glass." Everyone is familiar with the association of falling barometric pressure

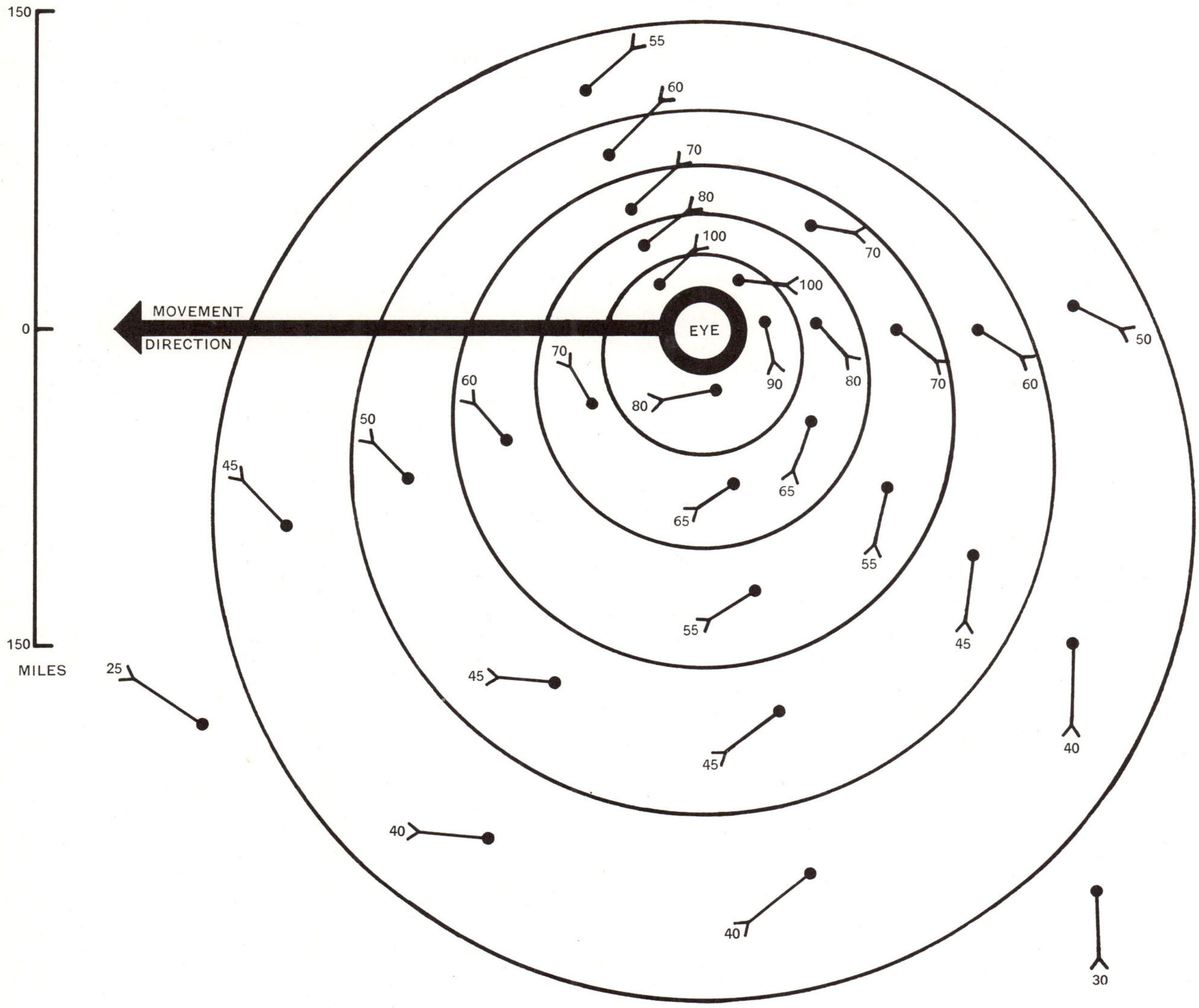

Fig. 14 Wind circulation around a typical 100-knot hurricane.

to oncoming stormy weather. Knowledge of barometric pressure patterns and their variations with time is the forecaster's most valuable tool.

Wind results from differences in barometric pressure between two points, somewhat as the flow of fluid through a pipe is governed by the pressure difference between the ends of the pipe. The pressure effect is in the direction of high to low pressure, and along the most direct route. This route is the one along which the pressure is changing most rapidly, and it is called the pressure gradient. If the pressure gradient forces were the only ones acting on the air, the wind would blow along AB in Figure 15, and it would accelerate because the pressure gradient is increasing in that direction. However, there are other forces acting upon the air. The Coriolis force—caused by rotation of the earth—always acts at right angles to the direction of air motion in the Northern Hemisphere. This force is proportional to the wind speed and increases in magnitude with increasing latitude. If the pressure gradient and Coriolis forces were the only ones acting, the wind would blow parallel to the isobars, CD in Figure 15. A second force, friction, acts opposite to the direction of air motion, retarding the flow. This reduces the magnitude of the Coriolis force, and the pressure gradient becomes dominant. The air then tends to flow across the isobars, from high to low pressure, EF in Figure 15. Roughness of the terrain determines the amount of friction and, in turn, the variation in the angle between wind and isobars. A third force of importance in hurricanes is the centrifugal, which opposes the pressure gradient. It becomes very large near the center of a hurricane and tends to hold the wind speed below what it would be if the pressure gradient had full effect.

Figure 14 has illustrated how the wind reacts to the several forces in forming a hurricane circulation.

Obviously, the hurricane must have a strong pressure gradient to have such strong winds. Since the hurricane is really not large as compared with mid-latitude storms, the central pressure necessarily must be extremely low.

The steepest pressure gradients in the world (barring tornadoes) are found during the immature stages of tropical cyclones in lower latitudes. One ship in the Caribbean experienced a pressure fall of 1.34 inches (40.5 millibars) in twenty minutes. Calculations made later showed that the pressure gradient between the fringes of the storm and the edge of the eye was 2.10 inches (70.1 millibars) in nineteen miles or 0.11 inches per mile. In the 1935 Florida Keys storm, one estimate gave a pressure gradient of one inch in six miles.

A typical barograph trace, showing pressure change during passage of a hurricane, is shown in Figure 16. If, like the ship with the bottom dropping out of the glass, your barometer is on the steep slope of the trace, you are in trouble.

The lowest sea level barometric pressure readings for the whole world have been recorded in hurricanes and typhoons. In the United States, the record is at Matecumbe Key, Florida, with 26.35 inches. The lowest accepted reading from a ship in the Caribbean is 27.01 inches. Undoubtedly, there have been many lower readings to be had for the taking, but who wants to set a record this way? Ships dodge the centers if at all possible.

The Eye

Surely the eye of the hurricane is one of our more curious meteorological phenomena. It is unique; no other atmospheric feature even comes close to a resemblance. Adding to the eye's aura of mystery is the fact that so few people have had the dubious advantage of seeing one.

What are the main features of the eye of the hurricane?

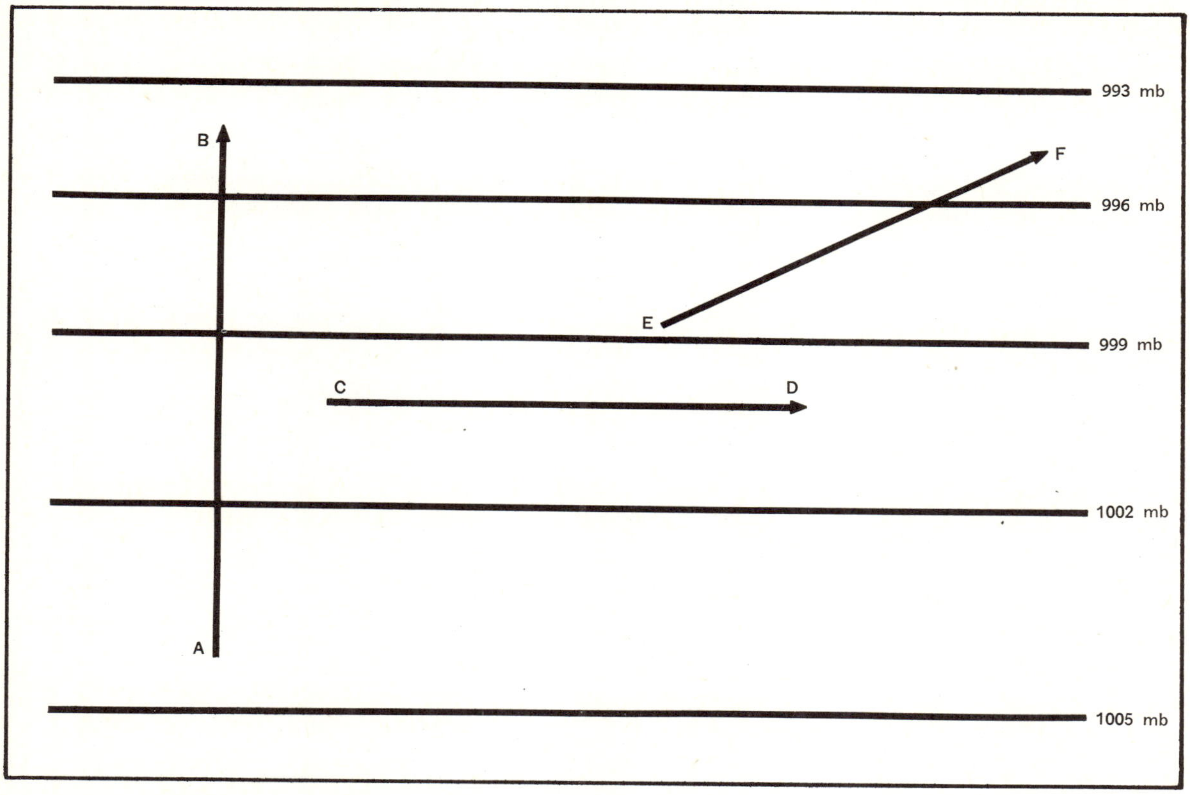

Fig. 15 *Pressure gradient force along AB. Pressure gradient and Coriolis forces along CD. Pressure gradient and Coriolis forces with friction added along EF.*

Fig. 16 *Barograph trace—Martinique, 8-9 August 1903.*

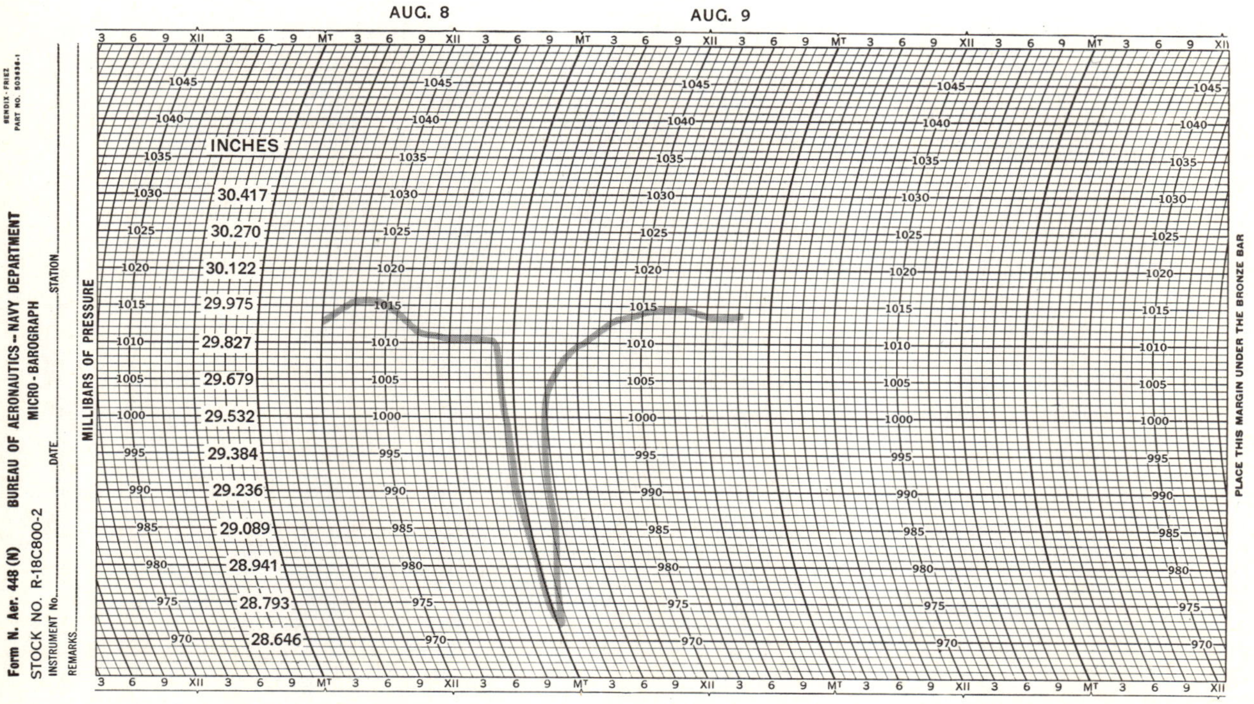

The full fury and strength of the hurricane's winds are found at the edge of the eye. As you cross this wall of maximum winds and pass into the eye, just like closing the door on a windy day, you find yourself in an area of nearly calm winds. The contrast is miraculous. You may go from 130 knots (150 mph) to 10 knots in a distance of 100 yards. Only the waves are worse, or seem so because the winds are no longer blowing off their tops.

In the eye, the dense, dark clouds are absent. Although there are usually some low clouds in the eye, if you fly above these you look up into bright sunshine, or perhaps see lazy-looking clouds near the top of the eye at 30,000 feet or more. You may even see a flock of tropical birds trapped in the eye and on their way to startle some New England bird watcher. Dr. R. H. Simpson (U. S. Weather Bureau) has made many flights into tropical cyclones. The following is his vivid description of the entrance into the eye of a storm:

> The plane flew through bursts of torrential rain and several turbulent bumps. Then, suddenly, we were in the dazzling sunlight and bright blue sky. Around us was an awesome display. Marge's eye was a clear space 40 miles in diameter surrounded by a coliseum of clouds whose walls on one side rose vertically and on the other were banked like galleries in a great opera house. The upper rim, about 35,000 feet high, was rounded off smoothly against a background of blue sky.

The average diameter of the wind eye is about 14 miles, but the range can be from 4 to 30 miles. Sometimes, the eye assumes an elongated shape along the direction of movement, and it has been reported as from 70 to 90 miles in diameter. The length of time it takes for an eye to pass a given location varies with the size of the eye, speed of the storm, and the portion of the eye which passes over the location.

Double eyes (actually concentric) have been reported occasionally, usually during some convulsive readjustment in a strong storm's intensity or in the size of the eye. A Navy meteorologist, Lieutenant Commander A. R. Fields, was the first to give a good description of this phenomenon:

> During the hurricane flight of the 13th, it was soon apparent that the storm had spread and intensified throughout the night. 40-knot winds commenced at a 100-mile radius, and moderate to severe rain was encountered from that point in. Turbulence was not severe, however, and the surface wind had not exceeded 80 knots when the aircraft, on a north heading at an altitude of 800 feet, emerged from a wall of nimbostratus into a zone of perfectly smooth air. There was a layer of broken stratocumulus overcast through which the sun could be seen as a round white ball. The wall of nimbostratus curved away to right and left, forming a perfect bowl, about 30 miles across. In the center of this bowl rose a column of vaporous, ragged clouds about 8 miles in diameter. As the plane emerged into the bowl, the weatherman remarked that if it weren't for the wind on the surface, he'd swear it was the eye. At that instant, the radar operator reported a perfect eye 5 miles dead ahead. It was perceived that the core must be the true eye of the storm. A series of hard jolts was felt as the plane entered the eye and seconds later, the plane emerged into a calm, hazy eye, remarkably similar to the eye of the previous day, with confused seas, and no wind at the surface. The pilot started easing to the left, intending to circle to the right inside the eye, when a line of foam was seen at the surface and it was realized that it was too late to turn. The left wing of the aircraft hit the wall of the eye first and was immediately flipped up. The plane dropped from 500 to 200 feet altitude before the momentum carried it into the wall of wind; then it started an uncontrollable 45-degree climb through severe turbulence. The pilots struggled to turn to a west heading and a few seconds later the plane broke out into the calm outer eye at an altitude of 2300 feet. . . . The remainder of the flight was routine.

Waves, Swells, and Surges

The state of the sea that results from hurricane winds, while not an atmospheric hurricane characteristic, is a standard feature of equal or greater importance than the winds which cause the state of the sea. A hurricane can be thought of as a moving area of heavy seas of some 40,000 square miles, an area within which ships must reduce speed to avoid damage.

Everyone knows that air moving across a water surface creates waves; blowing on coffee in a cup is a simple enough example. How fast the wind blows, how long it blows, and how far it blows are all factors which contribute to the ultimate height of the waves it creates. As the waves move away from the area of birth, they decay, become smoother, and are called swells.

The strong winds of tropical cyclones generate some of the highest of ocean waves. In an average hurricane, waves of 35 to 40 feet are not uncommon; in giant storms, they build to 45 or 50 feet. Waves still higher than this have been reported, particularly in the Pacific. Such reports are rare, and those waves in the 60- to 90-foot class are subject to considerable scientific doubt. Even a 40-foot wave, viewed from the deck of a ship, will make the most experienced observer uneasy and inclined to think he is seeing more than he actually is.

The highest waves are produced on the right side of the hurricane center (right with relation to direction of storm motion). This happens because the stronger winds are found in this part of the storm. Also, the winds have a longer time in which to push upon the water, since waves and storm are moving in the same direction. Figure 17 shows the way in which waves and swells move out from the storm center.

Waves move with a speed slower than the winds which create them, but still much faster than the movement of the hurricane itself. If a hurricane were moving at 15 knots and the winds on the "strong" side were blowing at 100 knots, the waves created would travel at 40 to 50 knots, and would soon outdistance the wind system of the hurricane. As they moved away from the storm, the waves would become swells, and continue onward ahead of the hurricane for hundreds or even thousands of miles. A characteristic of hurricane-generated swells is their long period; they arrive at the beach at a rate of about 2 to 4 per minute, contrasted with the normal 10 to 15 per minute. This feature, along with the gradual rise in water level from the hurricane swells, has long been one of the warning signs of the onset of a hurricane.

The hurricane surge, a sudden rise in water level, is associated with arrival of the hurricane and its strong winds at the coast. The surge, hurricane tide, and other effects from the state of the sea will be discussed in detail as forces of destruction.

The waves created near a hurricane center, as shown in Figure 17, move out in all directions because the creating winds are blowing in a circle. Consequently, there tends to be much crisscrossing of waves from different parts of the storm, creating the characteristically chaotic sea, in which waves seem to be going every which way with no sense of order or respect for the seamanship of good sailors.

Clouds and Precipitation

In the old days, clouds were the key forecasting means for tropical cyclones. Much was written, or passed by word of mouth, about the cloud sequence which indicated it was time to warn of the approach of a hurricane. The missionary priests were leaders in this activity, men like Father Depperman in the Philippines and Father Viñes in Cuba. Of course, these methods did not lead to early warning, because if the high, hurricane clouds are in sight, the storm center is comparatively nearby.

The precursory signs follow a remarkably uniform pattern. The long, heavy swell arrives well in advance of any hurricane clouds. The first clouds directly connected with the hurricane circulation

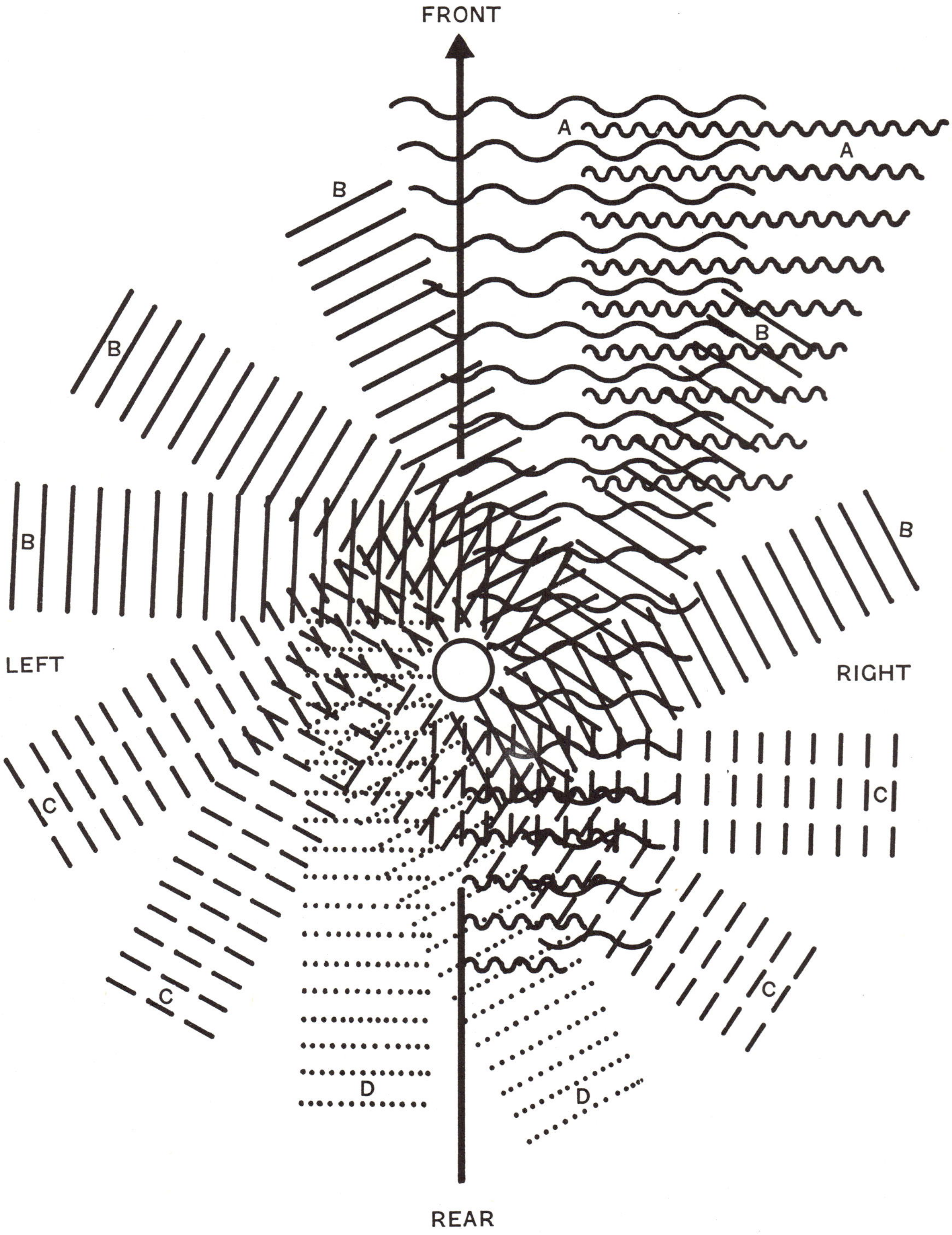

Fig. 17 *Generation of swells by a hurricane. (After Tannehill.) (From* Atlantic Hurricanes, *by Gordon E. Dunn and Banner I. Miller. Reprinted by permission of Louisiana State University Press.)*

A—Swells of greatest length and magnitude, traveling in line of advance of the tropical cyclone.

B—Swells of moderate length and magnitude moving to left and right of line of advance.

C—Smaller length and lesser magnitude moving outward to right and left.

D—Smallest length and least magnitude moving outward from rear.

are cumulonimbus (thunderstorm). There are active bands of these along a line several hundred miles ahead of the storm; the distance varies with the size of the storm, of course. Following these, and the day before the storm arrives, the tropical pattern is out of phase. The thunderstorms of the day before are missing, the usual cumulus clouds are suppressed, and there are bright skies and above-normal temperatures. Then, the barometer starts to drop, and the wind may come from an unusual direction. Next day, the standard hurricane cloud pattern starts.

First come the cirrus, the feathery ice clouds at the top of cloudland; the cirrus family includes the familiar "mare's-tails." These clouds seem to converge in the direction from which the storm is approaching, a characteristic most noticeable at sunrise and sunset. The cirrus clouds are followed by cirrostratus, the clouds next to the highest. These sometimes cover the whole sky like a veil, giving it a milky appearance. Often the cirrostratus clouds produce halos around the sun or moon, halos which have been successfully used as forecasting tools by laymen for thousands of years. The cirrostratus lower and change to altostratus. This cloud form resembles the cirrostratus in its fibrous, veiled appearance, but does not have the halo phenomenon. Mixed with the altostratus usually are the altocumulus, clouds at the same elevation but formed in globular masses rather than veiled sheets. After this come the cumulus congestus, the fulsome, bustin'-out-all-over looking beauties, with their heavy showers. The barometer falls more rapidly; the wind increases; finally, an ominous black wall of clouds approaches —the "bar" of the storm.

The hurricane is not the record breaker on rainfall, but it ranks close to the top. A hurricane's rain is always heavy, but it is difficult to measure the fall accurately. When the winds are over 50 knots, the rains of the hurricane blow almost parallel to the ground, whereas the rain gauge points toward the sky. Consequently, much of the rain beats against the side of the rain gauge and falls to the ground unmeasured. If a hurricane logs 5 inches at one spot, the actual fall is more likely to have been 7 to 9 inches.

L. A. Hughes (U. S. Weather Bureau) calculated that the average hurricane should have a fall of about 11 inches at any particular location over water. If the storm passes over land, the picture is distorted by the lifting of air by the land mass. It is over land that tremendous amounts of tropical cyclone rainfall are recorded, like the 49 inches in 24 hours at Paishih, Taiwan; the 96.5 inches in 4 days in Jamaica; and the many instances of falls over 20 inches in a 48-hour period at various coastal locations in the United States.

The rate of rainfall varies within wide limits. An exceedingly intense storm may log no more than an exceptional .25 inch an hour; other storms may have rainfalls of as much as 1.32 inches in 10 minutes, or 6 inches in 75 minutes.

Two other features of the precipitation-cloud structure of the hurricane are noteworthy. First is the thunderstorm, a familiar phenomenon in the middle latitudes. In the hurricane, it sometimes seems as though all the thunderstorms in the whole Mississippi Valley have been rolled into one. In hurricanes both at sea and ashore, there have been reports of lightning so frequent that the illumination seemed to be nearly constant. In one hurricane, reconnaissance aircraft flying near the eye were struck by lightning on three separate occasions. Thunderstorms are more commonly associated with the early stages of the hurricane, and the hurricane hunter crews have consistently reported that flights into the immature storms have often been rougher because of thunderstorm turbulence than flights in the giant-sized hurricanes. Second, tornadoes occur with a moderate frequency in hurricanes, but apparently only over land areas. Normally, it is impossible to see the customary terrifying clouds of the tornado because of all the other low clouds and the blowing rain of the hurricane. Tornadoes in hurricanes are ordinarily reported on the basis of the damage they do.

Dissipation

The tropical cyclone is the most persistent of all storms. Although some weaklings last only a day or two, the average life of hurricanes is nine days. Many have lived for three to four weeks, and the Methuselah of all lasted for five weeks. Meteorologists tracked it all the way from the African coast, across Puerto Rico and the Bahamas, to the vicinity of Cape Hatteras, and it could last be recognized as a tropical storm near the Azores.

The birth and mature stages of the hurricane have been examined, and now the death throes need a word.

If the center of a hurricane passes inland along the continental coast, the end is near. Friction between the wind and the rough land surface causes the winds to angle inward more toward the storm center— literally filling up the hole in the middle. A second aspect is that the storm is no longer fed by the life-giving flow of warm, moist air from the tropical sea surface. Usually, within twelve hours after a storm passes inland, the intensity has diminished 40 to 50 per cent, and within twenty-four hours, hurricane warnings are stopped and the center is reported only as an extra-tropical cyclone. Fast moving storms tend to hold their tropical character longer than the slow-movers. Some have entered the coast near Cape Hatteras, and racing to the north at high speeds, were still acting like tropical storms as far north as New York State.

Hurricanes which do not go ashore, but which continue northward over the Atlantic, meet their death because of passage over cold water and the entry of cold air into the storm circulation. As long as the storm can stay over the Gulf Stream, or get a "shot in the arm" from a surface cold front or approaching trough aloft, it tends to hang on to hurricane intensity. Otherwise it dies quickly, sometimes in less than a day. Ships sailing in northern waters must be as alert to hurricane warnings as those sailing in the tropics. Hurricanes have reached New Brunswick and Newfoundland with winds of 75 to 95 knots.

Section 4 : The Killer Side of Hurricanes

Hurricanes pack plenty of punch, and they should, considering the amount of energy involved. In one day, the average hurricane releases an amount of heat which, if converted to electrical energy, would provide a six months' supply for the whole United States. In a day, the hurricane will squeeze out twenty billion tons of water, the energy equivalent of half a million atomic bombs. Curious weathermen have made other types of energy calculations on the hurricane; all show the same thing—it is a tough customer.

The big killers can happen at any time. Hurricane Flora, in 1963, was next to the worst storm since Columbus came to the New World; an estimated 7186 lives were lost in Haiti, Cuba, and the Dominican Republic, plus a few more elsewhere.

The hurricane dispenses death and destruction in a variety of efficient ways. Although it would be natural to think of the wind as the most dangerous element of the storm, the actual order of importance in sorting out hurricane trouble is sea action, floods, and then wind.

Tides and Surges

Contributions from the forerunning swell, storm surge or tide, atmospheric pressure effects, and resurgencies all work to raise the water level at coastal locations. The effect of each varies greatly with the shape of the coastline, wind speed in the hurricane, speed of hurricane movement, and angle at which the storm approaches landfall. But, together, they produce the hurricane tide, the greatest killer in the tropical cyclone arsenal.

Swells from the waves of the hurricane move out well ahead of the storm center, sometimes a thousand miles or more in advance of a slow-mover. As the swells reach shallow water near the coastline, they cause a gradual rise in the water level. The forerunner effect, which may account for an increase of 2 to 4 feet above normal tides, occurs along several hundred miles of coastline.

The storm surge, a rapid rise in the water level, combines the effect of strong winds and barometric pressure falls. The peak of high water from the surge normally reaches the coastline about the same time as the hurricane center; the peak is just to the right (10 to 12 miles) of the point of hurricane entry, with gradually decreasing water heights outward 30 to 40 miles on either side of the peak. Fortunately, the hurricane surge affects only a comparatively small section of the coastline.

The barometric pressure-fall effect in the hurricane surge arises from the decreased pressure at the center of the storm, just less air pushing against the water. A one-inch fall in barometric pressure at a cyclone center allows the water level to rise one foot, so the maximum possible effect in hurricanes would be three to four feet.

The wind effect in the hurricane surge necessarily relates directly to the strength of the winds in the storm. The contours and height of the coastline, depth of offshore water, and angle of approach of the storm to land are other factors affecting height of the surge. The wind effect, simply stated, is just the piling up of water on the beach by the pushing force of the high winds; the water level reaches and maintains for a short time a certain height above the normal water level.

Hurricane tide or storm surge heights vary all the way from a few feet to as much as 20 feet, and this does not count the regular storm waves on top of the surge which crest much above the mean height of the storm surge. Surges penetrate 10 to 20 miles inland, depending upon the height above sea level of the land area involved.

Still another important effect is the hurricane wave. This is one huge wave among the regular waves superimposed upon the storm surge. It is a rare phenomenon, and reliable reports on it are few. It may just be a very rapidly developing or abnormal storm surge; some consider it possible that the hurricane wave arises from a rare combination of circumstances where everything—wind, water depth, waves— happens to be in perfect resonance to form a giant wave form among the already large waves of the hurricane. Some of the worst hurricane disasters, from the standpoint of loss of life, have been attributed to giant hurricane waves. The Galveston hurricane in 1900, in which over 6000 people drowned, had reports of a hurricane wave, as did the next two ranking hurricane killers. In India, the same phenomenon caused fantastic loss of life in past centuries—300,000 in one storm, 100,000 in another.

Resurgencies contribute to hurricane surge only under limited circumstances, but are of enough importance in other respects to be included. Resurgencies are created on the left side of an advancing hurricane, where the winds are blowing offshore. If the storm enters the coast so that the offshore winds blow across a bay or harbor, the winds will tend to empty that body of water. The water level will drop considerably, up to 15 to 20 feet in rare cases. After the center passes, the winds die down and the water rushes back to fill the hurricane-created vacuum. Then, the water rocks back and forth just like a bucket of water which has been tilted and then set back on a level surface. During this rocking process of the resurgency, the water level may reach several feet above the normal level at either end of bays and harbors. The effect varies greatly with the shape and size of the bay or harbor.

Waves

Waves in a hurricane are of concern both to the mariner and to the inhabitants of the shore lines. The development and motion of hurricane waves has been described, as well as the confusion in the sea pattern in a hurricane, with waves running in several directions, crossing, cresting at unexpected points, and making life miserable and dangerous for the unfortunate ship and its crew.

A ship caught in a hurricane is almost sure to suffer some damage, and it is possible the damage will be severe. At worst, she may go to the bottom to join the rotting, rusty hulks victimized by hurricanes of other years. In present times, owing to an improved warning service, most ships successfully avoid head-on encounters with hurricanes. Still, each year there are losses and heavy damage.

On shore, waves are responsible for much of the water damage. People, caught unaware by the hurricane surge, may drown, but most well-constructed buildings will stand up to the surge itself. It does not pound or beat things to pieces; it just covers everything with water up to the depth of the surge. The buildings can withstand this, but on top of the surge, huge and terrifying waves of the hurricane come roaring onto the beach at 50 to 60 mph. The force they exert is tremendous. A cubic foot of water weighs 64 pounds, a cubic yard 1728 pounds; the cubic footage which would be included in the space occupied by an automobile would weigh about 16 tons. Most readers have seen the damage that can be done by a 4000-pound car traveling at 50 to 60 mph. Visualize, then, the effect of over 30,000 pounds hitting at the same speed. Buildings are not designed for this sort of treatment. Consequently, whole communities have been wiped out when a surge has covered the city with water and given the waves a chance to wreak their destruction.

Sometimes, the waves do not have to get an open shot at the buildings; it is enough for them to work on the ground underneath. Buildings may be undermined quickly, and down they come. This type of destructive force is shown most vividly along the shore line, where damage to beaches is nearly always pronounced during hurricane passage. A full-blown hurricane can dig out 40 feet of beach in short order.

Floods

Floods are second on the list in dealing out death, injury, and destruction. Hurricane Diane, in 1955, often called the "first billion dollar hurricane," earned most of its bad reputation because of floods. Diane left in its wake about 200 deaths, and damage estimated at over three-quarters of a billion dollars; most deaths and damage came from floods. There may never again be the same combination of circumstances which were present with Hurricane Diane. It followed right on the heels of Hurricane Connie, and found the ground soaked and streams already full when it started dumping rain in amounts of ten inches and more in twenty-four hours from Virginia northward. However, there will be no shortage of hurricane floods; there has been an average of more than one every other year in the United States. Nearly every coastal state has been affected at one time or another, and even some inland states such as Oklahoma, Ohio, Missouri, and West Virginia have had their turns with hurricane-induced floods.

You are likely to be on your own to a greater degree so far as floods are concerned than you are with the hurricane movement or storm tides. The latter can be forecast with a fair degree of accuracy; however, river forecasters must have some idea as to the amount of rainfall to be expected over the watershed involved in order to issue a proper forecast. To date, there has been no satisfactory method developed for forecasting rate of rainfall. So, with the extremely high rate of rainfall possible in hurricanes, the river forecaster may still be working on his first guess when the rivers overflow.

Winds

The snarling fury and powerful force of strong hurricane winds cannot be adequately described. You learn about them only by experience—an experience never to be forgotten. You can feel the power of the wind wherever you are. At sea, you can feel the racked and groaning ship shudder as it fights against being rolled over; ashore, not only does a building shake, but it also seems to twist its whole frame; outside, you must lean 20 to 30 degrees into the wind to keep from being blown over. You can hear the wind wherever you are—it howls; it screams; it roars. It attacks you with pounding rain, sea water, leaves, boards, anything loose against your shelter. It can drive a board through a thick tree. It can sandblast a man to death, leaving him with no skin and no clothes except his shoes and belt.

The wind velocities to be found in hurricanes have been given—90 knots as an average, maximum up to 150 knots in most severe storms. The fact has been stressed that gusts may reach up to 50 per cent over the basic wind speed; with 150-mph winds, gusts may reach to 225 mph or more. This wrenching, rocking effect from the gusts must be considered carefully in building design.

The wind increases its force upon a solid object with the square of the velocity. A wind that doubles its speed increases its force four times. A 60-mph wind exerts 15 pounds per square foot pressure; a 125-mph wind exerts 78 pounds; a 150-mph wind, 112 pounds.

Appreciable damage occurs in any inhabited area when the pressure per square foot reaches 15 to 20 pounds, but this is particularly true in tropical countries where the economy of many areas dictates poor construction. Whole cities have gone down, some never to be rebuilt. This was the case in Chetumal, Mexico, where Hurricane Janet, in 1955, left only four battered buildings in this community of 2500 people. In 1961, Belize, British Honduras, suffered a similar fate.

Section 5 : The Hurricane Warning Service

At sea or ashore, when a hurricane is on the prowl, warnings are issued every six hours; if new information requires a change in the warning, bulletins are issued at shorter intervals. Who prepares these warnings; what information has been used; can you depend upon these warnings?

Collection of Weather Information

The first job of the hurricane warning service is to collect observational information on the weather, not only from tropical areas but also from northern latitudes. Particular attention must be paid to those areas near incipient or active tropical cyclones.

Surface reports are collected from all the major land masses surrounding the hurricane belt. From islands in the Antilles, Bermuda, Azores, and Cape Verde regular 6-hourly surface reports are received. Many of these same stations provide upper air reports, which give information as to winds, temperature, pressure, and humidity at levels between surface to about 100,000 feet.

From the hurricane spawning grounds, ship weather reports are available. Although there are not many of them, these reports are of great importance. About half of all tropical storms are detected initially on the basis of ship observations. Be sure your observation is sent to your normal weather report collection center.

Both island and ship reports are relayed every six hours, although reports may be requested from suspicious areas as often as every two hours.

Ship and land station weather reports are absolutely essential to the hurricane warning service, but developing tropical storms are not always considerate enough to be located near a land station or a ship. In a fully developed hurricane, the number of ship reports usually dwindles to nothing, because the ships do their best to get as far away from the storm center as possible. If the warning centers are to do much more than guess, they must have another source of information. The hurricane hunters, of which so much has been heard in the last two decades, fill the information gap.

The conditions favoring tropical cyclone development have been outlined, and some of the weather signs for which to watch have been given. Hurricane forecasters watch for the same signs from their ship and island weather reports, and when enough evidence accumulates, a reconnaissance plane is sent out to examine the area for storm development. The plane may find nothing definite, only a suspicious-looking situation. Sometimes, flights are made into one of these areas for as long as a week before the storm develops, if it does.

Once a tropical storm or hurricane is uncovered, the hurricane hunter planes give it almost a 24-hour-per-day examination, sifting out all the facts on surface and upper winds, state of the sea, pressure, and the eye and storm location. Without the reconnaissance information, the forecast centers would not be able to do any better than they did thirty years ago. Small storms would be missed for days or forever; intensities would be misjudged; movement forecasts would be highly inaccurate. Hurricane reconnaissance is a difficult, dangerous job, but one that has saved many lives and prevented much damage. In a very active hurricane season, with about twenty tropical cyclones, the reconnaissance job will require approximately 275 flights for a total of a little over 3000 flight hours. And these are rough hours.

Radar and weather satellites are able to provide useful information as to hurricane location, and hurricane intensity to some degree, but they are far from a stage of development where they might replace aircraft reconnaissance. Satellites, in many cases, have uncovered unsuspected storms in remote areas, and they are able to track these storms with a fair degree of accuracy. Methods for determining storm intensity from satellite pictures are under development, and within a few years, estimates of wind speed in hurricanes should be accurate and reliable enough for operational use. There have been several cases where the meteorological satellite has outlined a cloud pattern which could be interpreted as a tropical cyclone—but there were no winds. Further experience in interpretation of satellite pictures should soon eliminate this type of error.

Long-range radar can locate storm centers with accuracy about half the time; in the remaining time, there is difficulty with tracking because of the indefinite structure of the eye—either it is not formed with well-defined wall clouds, or it is in a momentary stage of change. Radar can give some indication of intensity, but is subject to a wide range of error. Generally, if there is a partial wall cloud, with long, strong feeder bands, the winds are in the 65- to 75-knot range. If the eye is completely closed, with many feeder bands, the hurricane is usually in the 100- to 135-knot class.

In general, radar and satellites are extremely useful, but are complementary to aircraft reconnaissance. Radar and satellites, particularly the latter, are the bird dogs of the modern hurricane hunting system. They tell the forecasters where to send the planes to locate incipient storms, thereby saving useless flights. With full-grown storms, they tell the plane crews the most direct track to the storm center.

Warning Preparation

At the warning centers, information from every possible source is collected, plotted, and analyzed by a group of men with special training and long experience in tropical meteorology. Every modern method, both objective and subjective, is used in the forecast preparation. Considering the limitations imposed by lack of data and the many gaps in our knowledge of the dynamics of tropical cyclones, the forecasts are surprisingly accurate.

A major improvement in forecasting accuracy followed when upper air charts became available in the 1930's, and another when aircraft reconnaissance was started in the mid-1940's. Since then, there has been a slow but steady improvement in forecasting accuracy. With the research now in progress, further improvement can be expected to the point where the average 24-hour forecast position error on hurricanes will be less than 100 miles. Increased understanding of tropical cyclones, because of active research, should have its greatest effect upon the forecasters' ability to deal with storms with unusual movement and character.

There seems to be little chance of improving hurricane forecasts to the point where there is pinpoint accuracy on movement forecasts—say within 10 to 20 miles for 24 hours ahead. It is just not economically feasible to gather the data necessary to make such pinpoint accuracy possible.

In planning, then, you will have to live with and operate under average 24-hour forecast errors in the 100- to 120-mile bracket, knowing that there will be some better, some worse. After all, a 100-mile error is not too bad for 24 hours. A 5-knot error in forecast speed alone would exceed this, and if you throw in only a 2- to 3-knot error and add a 10-degree direction error, you will make up your 100 miles.

Before leaving the forecasters, and in view of the small number who are highly qualified, a recommendation must be made against the Chinese method of dealing with their tropical weather experts, even though it might leave a residue of nothing but the finest. A 1964 newspaper item is quoted:

> TAIPEI, Formosa, Jan. 8—The Control Yuan, top watchdog body in Nationalist China, has impeached the chief of the weather bureau for dereliction of duty.
>
> . . . was accused of having provided inaccurate information about typhoon Gloria which struck Formosa . . . leaving 239 persons dead and 89 missing.
>
> The Control Yuan said Mr.____'s information that the typhoon had changed course and would not hit Formosa caused people to neglect precautionary measures.

Warning Distribution

Hurricane warnings are distributed rapidly and widely; everyone concerned should receive the information, except for a few who may not have access to radio or telephone. Special teletype circuits link the Gulf and East Coast states, so within minutes after the harried forecaster writes his final words of advice, nearly everyone has received them. From central offices, warnings are relayed to radio and TV stations. Ships and small craft receive warnings over marine broadcasts.

Warnings—Content and Evaluation

The terminology and content of a typical hurricane warning follows, and for operational use, an accuracy evaluation is included:

Sample Warning	Content and Evaluation
Miami Weather Bureau Advisory Number 20 Hurricane Sally 11 AM EST Friday October 20 1964	*Hurricane advisories are numbered to ensure that you are reading the right one in a sequence. Dates and times occasionally get garbled in transmission, so the advisory number serves as a check.*
Hurricane warnings are displayed from Charleston South Carolina to Cape Fear North Carolina and people in this area should continue their preparations for hurricane force winds and abnormally high tides.	*The display of hurricane warnings signifies that within 24 hours winds of hurricane force are to be expected within the display area.*
A hurricane watch is in effect and gale warnings are displayed north of Cape Fear to and including Cape Hatteras. Gale winds are now occurring just off-shore along the Georgia and South Carolina coasts and are expected to move onto the coastal areas of South Carolina today and begin to affect the North Carolina coast late tonight or Saturday. Small craft from Georgia to Cape Hatteras should remain in safe harbor.	*A hurricane watch means that there is a possibility of hurricane winds reaching your area sometime beyond the next 36-hour period. The "hurricane watch" means that you should make* preliminary *preparations, and be prepared to make* final *preparations with no more than 24 hours' notice. The term "gale warning" means that winds in the 34- to 64-knot range may be expected in the area of display within 24 hours.*

Tides are expected to be from 6 to 10 feet above normal occurring mainly in the area from Wilmington North Carolina to Myrtle Beach South Carolina where the center of the hurricane is most likely to enter the coast tonight, probably after midnight. Extensive flooding and beach erosion will occur in this area and all people living in low lying locations in the area of hurricane warning display, but especially from Cape Fear to Myrtle Beach, should move to higher ground.

Tide warnings of 6 to 10 feet mean that tides of this height are expected during the passage of the hurricane over the coast. The 6- to 10-foot range, in effect, means that the coastline contours vary and that the forecasters do not know exactly where the hurricane will go ashore. You should prepare for the higher tide height; and if you happen to be located near or on a closed bay or estuary, you should anticipate tidal heights greater than those given in the warning. This is particularly true at the end of a long estuary where the tides may be nearly twice the height encountered on an open beach.

Coastal radar and air reconnaissance locate the center of Hurricane Sally at 11 AM EST . . . 1600Z . . . near latitude 32.1 North longitude 79.4 West or about 65 miles southeast of Charleston South Carolina moving on a northerly course at 8 mph.

The accuracy of the hurricane center position when both radar and aircraft are combined should be within 10 miles. The only complication would be when the storm has a peculiarly shaped or partly formed eye; here, it is hard to tell where the center actually is. The warning will customarily make mention of this if necessary. If the storm position is based entirely on reconnaissance data, you may expect accuracy within 25 miles if the plane is navigating by dead reckoning, within 5 miles if the plane is flying near land and can get navigational fixes by radar, and within 10 miles if using LORAN (electronic navigator). If the storm position is based on ship reports, you may expect an error range of 10 to 50 miles. Ships normally tend to evacuate the area of a hurricane as fast as possible, and if they should be caught, their navigation errors are often extreme.

Highest winds are estimated 100 mph over a small area near the center and gales extend outward 175 miles over the ocean to the northeast and about 100 miles to the southwest of the center. The lowest sea level pressure this morning was 976 millibars or 28.82 inches.

A daylight center fix reported by reconnaissance aircraft means that the maximum winds and general wind pattern have also been given by the plane. If position and winds are based upon ship reports, an error of 25 knots or less may be assumed. Ships report from only one position for one warning, and there are not many ships reporting. If winds are "estimated," an average error of 30 knots should be allowed for, but it is possible to get extreme errors in rapidly developing or dissipating storms—sometimes as much as 50 knots.

Sally is forecast to move on a North Northeast course at 12 mph during the next 18 hours.

The average 24-hour forecast error is about 120 miles. The Weather Bureau, unless the storm is expected to go ashore, gives direction and speed forecasts rather than specific locations for advance periods; the latter type forecasts are to be found in those issued by the Navy and Air Force. The Weather Bureau movement forecast can be moved forward for, say, 24 hours. Average error with the end result will be 120 miles or more. With Navy forecasts, you can use the 120-mile error for operational purposes.

If you have worked with hurricanes at all, you know that the 120-mile average error means 120 miles in any direction from where the center is at the end of 24 hours. The storm may speed up; it may be right on track, but ahead of schedule; the final position may be fine for distance, but poor on direction. It might even be fine for distance, but have reversed itself and be headed in a direction opposite to that stated in the warning. Nevertheless, work with your 120-mile average error. As was pointed out in a previous section, this is still good forecasting.

No further intensification or change in size is likely.	*Accurate forecasting of intensity or change in size is one of the weaker links in the hurricane business. Some of the forecasting rules work well—but not consistently. The tendency among forecasters is to rely mostly on extrapolation. That is, if a storm is gradually getting weaker, the forecaster indicates that it will continue this trend. Often, in the early and late stages of hurricanes, pure extrapolation can lead to gross errors. Another thing to watch for is evidence of the well-recognized tendency for storms to have short-period fluctuations in intensity, with a change in the maximum winds covering a range of 25 knots in as short a period as 12 hours.*
Even though the center is expected to enter the coast probably after midnight near or just southwest of Cape Fear, all persons are cautioned that dangerous winds, increasing high tides, and heavy rain precede the center of the storm. All interests in the Carolinas should keep in close touch with all future advices. Shipping should exercise extreme caution.	*The statement relating to time of passage of the storm over the coast with preceding winds and tides is important. Many people hear a warning which says that Hurricane Somebody will arrive at a certain time and think they are comparatively safe until then. It would seem quite obvious—but apparently is not to many—that the first half of the storm is going to beat you to a pulp before the "eye" ever gets ashore. The area of gale and hurricane-force winds should be related to your location, not just the storm center alone.*
The next regular advisory will be issued by the Miami Weather Bureau at 5 P.M. EST. There will be an intermediate bulletin at 2 P.M. EST and all interests will receive statements on local conditions from their respective Weather Bureau offices.	*Intermediate bulletins usually have little of value in the way of new planning information, unless there has been additional reconnaissance information on position or storm intensity. Normally, there can be little change in the storm in the 3-hour period after issuance of the regular advisory.*

To show the variation with Weather Bureau warnings, a sample Navy hurricane warning is now quoted. These warnings are designed more for putting the decisions up to the operators; there is less verbiage included, as it is assumed that the Navy decision-makers want only the facts and that they will make their own decisions. Customarily, these decision-makers are men with long experience at sea, as well as with harbor and shore installations which serve the Fleet. The sample warning: ⇝

Heading (usually long, with lots of ADDEES)

BT

UNCLASSIFIED

FOR OFFICIAL USE ONLY

HURRICANE CLEO WARNING NUMBER TWO THREE. POSITION TWO TWO POINT ZERO NORTH SEVEN NINE POINT FOUR WEST AT TWO SIX ONE ZERO ZERO ZERO ZULU. POSITION GOOD BASED ON TWO SIX ZERO SIX ZERO TWO ZULU RECON FIX. PRESENT MOVEMENT THREE TWO FIVE DEGREES ZERO NINE KNOTS. MAXIMUM WINDS OF SIX FIVE KNOTS IN SMALL AREA NEAR CENTER. FIVE ZERO KNOT WINDS RADIUS FOUR ZERO MILES FROM CENTER IN ALL QUADRANTS. THREE FIVE KNOT WINDS RADIUS EIGHT ZERO MILES FROM CENTER IN ALL QUADRANTS. TWELVE HOUR FORECAST POSITION TWO THREE POINT FIVE NORTH EIGHT ZERO POINT FOUR WEST. MAXIMUM WINDS FIVE ZERO KNOTS. TWENTY FOUR HOUR FORECAST POSITION TWO FIVE POINT ZERO NORTH EIGHT ONE POINT TWO WEST. MAXIMUM WINDS SEVEN FIVE KNOTS. FORTY EIGHT HOUR OUTLOOK POSITION TWO EIGHT POINT ZERO NORTH EIGHT THREE POINT FIVE WEST. MAXIMUM WINDS EIGHT ZERO KNOTS. SEVENTY TWO HOUR OUTLOOK POSITION THREE ONE POINT ZERO NORTH EIGHT TWO POINT FIVE WEST. MAXIMUM WINDS FOUR ZERO KNOTS. NEXT WARNING TWO SIX ONE SIX ZERO ZERO ZULU.

BT

Section 6 : Tips from the Old Weatherman

A statement made by Fleet Admiral Chester W. Nimitz in 1944 is as meaningful today as it was then—and it will continue to have meaning in the future: "The time for taking all measures for a ship's safety is while still able to do so. Nothing is more dangerous than for a seaman to be grudging in taking precautions lest they turn out to be unnecessary. Safety at sea, for thousands of years, has depended on exactly the opposite philosophy." The Admiral's statement is equally applicable ashore.

It is beyond the scope of this book to give advice on every possible contingency, but certain useful, fundamental rules will be outlined for three situations: At Sea or Ashore, At Sea, Ashore.

At Sea or Ashore

1. Know thoroughly and have readily available all OpNav, CinCLantFlt, Operating Base, and Unit Instructions pertaining to warnings and conditions of readiness concerning destructive weather phenomena. These outline definite procedures to be followed in setting and following Heavy Weather Conditions. Some coastal cities have developed plans for action in case of hurricane threat. You should be familiar with these, and keep tuned in for the latest guidance from municipal officials. All of these instructions, both military and civilian, have been developed on the basis of sound knowledge and hard experience.

2. Keep a plot of the official warnings if there is any threat to your location. This will keep you alert to the need for preliminary planning, and ensure that you have the correct information. Be sure you have official warnings, and not some local interpretation.

3. If you are ashore and concerned with a decision on using or evacuating a harbor, or at sea and concerned with the problem of running or not running for safety within a harbor, the decision must be based on thorough knowledge of many factors: availability of moorings and docks, and length of ship as related thereto; depth of water at the piers; quantity and condition of its service craft; efficiency of port services; and how all these are affected by damaging winds, waves, and tides. Reference must be made to the latest hurricane information to determine the direction and force of the wind in the harbor, as well as the expected height of tides and waves. The Commander, SEVENTH FLEET described the result of such a decision—in relation to a typhoon—as follows: "Violent VIOLET, a large and vicious typhoon, . . . passed 30 miles southeast of Yokosuka 10 October. At Yokosuka most SEVENTH Fleet ships chose to ride her out in port. There was no damage to ships or significant interruption to the SEVENTH Fleet schedules." On this occasion, 37 ships elected to stay in port to face wind gusts forecast to reach 70 to 80 knots. The confident decision by the Senior Officer Present Afloat and the 37 skippers was based on sound knowledge of the harbor and refuge it provides.

4. If the decision has been made to stay in port, the following steps should be taken for securing for heavy weather:

a. Double normal mooring lines with one-inch wire lines doubled.
b. Run head line to shore, with slack to be tended if surging fore and aft occurs.
c. Where applicable to the berth, moor with bow to sea wall to keep screws clear for backing, should lines part.
d. Recall personnel.
e. Engineering plant on stand-by.
f. All holds re-shored and secured.
g. Sea detail with sound-powered phones standing by.
h. Heavy weather gear and flashlights broken out for use.
i. Rescue-Assistance lockers inventoried.
j. Dock area policed and loose gear secured.
k. Radio and signal watch maintained.
l. Continued mooring line inspection maintained.
m. Maintain plots of storm on bridge, in Combat Information Center (CIC), wardroom, and messdeck.
n. Hold briefings for crew on status of storm.

5. If the decision has been made to stay in port, but at anchor (or moored since the bridle or hammerlock moor will be used):

a. Take steps listed in paragraph 4, above, except for the parts pertaining to lines ashore.
b. Select a position with room to swing to a circle equal to the length of the ship plus the longest scope of chain used. Allow a safety margin for any dragging, by your ship or neighboring vessels.
c. Study the storm movement carefully to determine in which half you will be as the storm passes.
d. After the riding anchor (long-scope) is dropped, drop the short-scope, underfoot anchor on the side of the ship which will permit it to be dragged as the wind veers or backs so as to maintain an open "V" in the bridle moor and keep the ship heading into the wind.
e. Use main engines to offset the wind. The number of knots of engine speed necessary to do this will vary with the type of ship and loading, and with the wind speed. Ordinarily, with winds of hurricane force, experience shows that it will be from 3 to 5 knots.
f. Keep close watch for signs of dragging by using one or more of the following methods: bearings of objects on shore, radar, drift lead, and continuous echo soundings. The riding chain should be watched continuously for the characteristic jumps caused by a dragging anchor.

6. Make full allowance in your planning for the average forecast errors discussed in preceding sections.

7. Be sure your radar gear, if you have it, is ready for action when the storm gets near. Pinpoint knowledge of the storm's position on a continuing basis can be used to great advantage in ship maneuvering and in determining the specific area which will be most heavily affected by high winds and waters.

8. If you have no official warnings, but note an unusually long swell with a frequency of 4 to 5 per minute as compared with the usual 12 to 15, you can be sure there is a well-developed tropical cyclone in existence located in the general direction from which the swells are arriving.

9. If the surface pressure drops below 1000 millibars (29.53 inches) at your location, particularly if the rate of fall exceeds the normal diurnal change, make preliminary preparations for a storm. The

weather centers may not have the word yet. Repeat your weather report and seek advice from the centers.

10. If the eye passes over your location ashore, stay where you are until the second half of the storm goes by. At sea, all you have in this situation is the meager comfort of moving easily into the "safe" semicircle.

11. If you have had no warnings, but have seen the unusual swell and start to see the typical cloud sequence associated with hurricanes, break into full action to maneuver and secure.

At Sea

1. Determine the bearing, distance, and track of the storm from the official warnings, or from your own calculations if there are no warnings. From this information, you can plan how best to avoid the dangerous semicircle of the storm. Relationship to shoal water must be considered in the planning.

2. If you are near a storm and have no warnings, determine the bearing of the storm by the direction from which swells are arriving and by adding 115° to the direction from which your observed true wind is blowing.

3. If the wind gradually hauls to the right (clockwise), the ship is in the dangerous semicircle, *i.e.*, the right side of the hurricane in relation to storm track. If it hauls to the left (counterclockwise), you are in the safe or navigable semicircle.

4. If the wind remains steady in direction, increases in speed, and the barometer continues to fall, you are directly in the path of the storm.

5. Use your radar.

6. Do not try to outrun or cross the "t" of a hurricane; it usually means trouble. The main difficulty arises from the front-running swells of the hurricane, which build rapidly in size with the approach of the center. These can cut down your speed by several knots, while the hurricane keeps on at its own appointed pace—or speeds up.

7. If Sea Surface Temperature charts are available and your maneuvering plans permit, avoid the areas of warmest waters. Hurricanes like these as a path if their speed is slow, say 10 knots or less. If the storms are moving fast, 16 knots or more, the warmer waters have little influence.

8. If you are actually caught in the hurricane circulation, even the fringes, take these steps:

a. If you are dead ahead of the center, bring the wind on the starboard quarter (160° relative) and make best speed on this course. This will get you away from the center most quickly, and put you in the safe semicircle of the storm.

b. If in the safe or navigable semicircle, bring the wind on the starboard quarter (130° relative) and make best speed.

c. If in the dangerous semicircle, bring the wind on the starboard bow (45° relative) and make as much headway as possible.

d. If necessary to heave to, do so head to sea. There is an article on page 179 which proposes that all ships caught in a tropical cyclone shut down engines completely and drift with the storm. This seems to work well for some types of ships, but many mariners maintain that top-heavy men-of-war, as well as other ships fully loaded and of deep draft, would be in danger if this dead-in-the-water procedure were to be followed.

e. Proximity to land areas, coastlines, or islands must be taken into account in maneuvering. An evasive course which brings you close to shore is dangerous, particularly if you are in the sector of the storm with onshore winds.

Ashore

1. Make a thorough inspection of all buildings and other structures so that repairs and emergency bracing can be made well in advance. Means for securing all structures should be at hand and readily available for use if the need arises. Buildings should have been evaluated against standard hurricane building codes beforehand, in order to determine which buildings will need more than the usual anchoring and strengthening.

2. A thorough cleaning up of all loose material must be conducted. Chairs, tools, doghouses, garbage cans, boards, and toys can all be lethal weapons in a hurricane. Winds that will move Dempster Dumpsters and automobiles have no trouble with smaller items. Dead or dying trees should be cleared away.

3. Sterilize the bathtub and all other available containers large and small. Fill them with water in case the regular supply becomes contaminated.

4. Lay in a stock of canned and prepared foods, charcoal, Sterno, extra flashlight batteries and bulbs, matches, first aid supplies, and a few bottles of "snake bite medicine."

5. Have some extra lumber, wire, pliers, tools, and nails on hand for making emergency repairs.

6. Have a battery-powered radio on hand to keep up with what's happening to the hurricane and your neighbors.

7. Board-up or shutter-up early, so you do not have to do it in the wind and rain. Use good boards and secure them firmly. If one window gives way on the windward side, everything in the building will get soaked, and probably some of your other windows and doors will go with the wind. Leave a window or opening on the lee side which can be opened a little to equalize pressure in the building. Ordinarily, the pressure will equalize through small cracks and crevices, but a very tight building or a very rapidly moving storm may spell trouble. The Navy's hangars at Richmond, Florida, literally exploded during passage of a hurricane.

8. Set refrigerators and freezers to coldest setting. They will hold low temperatures for up to 48 hours after you have disconnected the power.

9. Fill the gas tanks in the cars. Gas pumps are driven electrically, and might be inoperable for a few days.

10. Disconnect main electrical switches and shut off the gas.

11. Evacuate coastal areas, particularly any that are likely to be flooded with a 20-foot rise in mean sea level. If on an island, leave early or you may not make the causeway. Keep in mind that evacuation may be slow, with clogged roads. In Hurricane Carla (1961), 500,000 people moved to high ground, creating a serious traffic problem.

12. Take small boats to a safe place, preferably a mile or more away from the ocean. If possible, take them out of the water and put them in a safe building, or secure them in a sheltered harbor with plenty of line fore and aft to allow for rising waters. Do this well ahead of time; many a small boat sailor has been lost to his dear ones trying to move his boat during a hurricane.

13. How to behave? First, keep calm. Hysteria spreads and always leads to poor judgment. If the eye goes by, a few emergency repairs might be in order, but be sure to get under cover before the back half arrives.

14. After the hurricane is over, get medical care as necessary; do not touch any loose or dangling wires but report them; be very careful in use of emergency lighting or cooking equipment so as to prevent fires; stay away from disaster areas unless qualified to help; walk and drive cautiously as trees may fall, bridges collapse, roads cave in; hose down shrubs and trees if they were sprayed with salt water; dry off metal which might rust.

Section 7 : The Future

Those who have read this far will surely stay out of trouble now whenever hurricanes are on the prowl in their vicinity, but a look at the future might also be in order.

A review of the elements of a hurricane warning and the comments made in earlier sections will show that the perfect hurricane forecast will probably never be realized. We can, however, look for improvements.

In 1954, Hurricane Carol, after moving lazily along the southeast coast of the United States, suddenly darted up the coast and arrived in New England, via Long Island, somewhat ahead of the time announced in the warnings. Surprise was not limited to the forecasters, but was shared by many worthy citizens in the New England area whose property was severely damaged. The ensuing anguished cries from New England were heard in Washington, and the weathermen were called on the carpet. They explained that there is not enough known about hurricanes to give pinpoint accuracy. The Congress thereupon appropriated funds for research in order that the forecasting of these destructive storms might be improved.

The National Hurricane Research Project was established in Miami in 1955, under Dr. R. H. Simpson of the U. S. Weather Bureau. This unit has achieved much of value since then in uncovering information about the tropics. Their work has led to a better understanding of the hurricane as an atmospheric phenomenon and to some improvement in forecasting accuracy. Further improvements can be expected: perhaps a major breakthrough in our understanding of how hurricanes form, what forces are involved in their movement, and how to employ this new knowledge in forecasting.

Computers

Another means by which forecasting accuracy will be improved is in the use of computers. The Navy and Weather Bureau have worked on the problem at their numerical centers, and have reached the point where their computer predictions approximate what might be expected of a topflight hurricane forecaster. The machines do their forecasting by the steering method which, simply stated, means that the hurricane is carried along by the air stream in which it is embedded. The computers perform calculations which "remove" the hurricane circulation from the carrying air, rebuild the pressure pattern without the storm, and then examine the air stream to answer the question of where something would go if it were to be steered by this flow. The winds considered are from surface to about 40,000 feet.

Experts in numerical forecasting believe that a 10 to 20 per cent forecasting improvement can be expected of the computers over the next few years. What limits the machines—just as it limits man—is a lack of data and equations which completely describe the atmosphere. There are just not enough reports concerning the hurricane to make a 100 per cent accurate forecast, and there never will be. It is not economically feasible to provide oceanic coverage equal to that existing over the continental United States.

Another facet of machine forecasting under development is that of predicting the areas of probable hurricane development. A considerable degree of success may be anticipated in this endeavor, since the computers, because of their speed of operation and global coverage, are better able to introduce high latitude effects upon the tropics.

Satellites, Radar, and Buoys

In the field of surveillance, improvements can be expected in three areas. The first is in meteorological satellites. Some of the difficulties involved with satellite hurricane detection have already been discussed: the satellite may report a cloud mass as a hurricane when it is not one; there is a relatively large degree of error possible in fixing the hurricane center position and determining intensity. However, in the next few years, added experience in interpretation, and possibly improved sensors aboard the satellite, will eliminate these weaknesses.

A second area of prospective improvement in surveillance is in radar. At present, it is restricted to ranges of 200 to 250 miles, the curvature of the earth causing the restriction, as radar beams travel in approximately straight lines. The researchers must find some way to bend the beams to fit the curvature of the earth and to build sensitive enough receivers to pick up the return signal. When this is done, and suspicious areas or hurricanes hundreds of miles out can be located, part of the burden now placed on hurricane reconnaissance aircraft will be lifted. However, the aircraft will still have to go out to verify the existence of trouble spots and to locate established circulations. The aircraft will also have to track the established storms, since the eye structure is so variable that radar cannot be depended upon much more than half the time.

A third area of potential improved surveillance is in weather-reporting buoys. The Navy has been developing these for several years, and they are about ready to play an important part in the weather-reporting network. The first NOMAD (Navy Oceanographic Meteorological Automatic Device) was tested in the Gulf of Mexico in 1960, and its reports were the first indication of Hurricane Ethel. The NOMAD relays selected weather elements to a shore station every six hours, or oftener if the wind should be stronger than 21 knots. A recent model is employing atomic energy for its power source. Within a few years, these anchored buoys will be providing meteorological and oceanographic information from many key points in the tropics. They will improve both forecasting and research.

Storm Modification

The one activity which has the greatest potential for affecting the hurricane is modification. This subject has been treated by the scientist, the pseudo-scientist, and the ignorant. Suggestions for hurricane modification number in the hundreds. At the Miami Hurricane Weather Central, it has been common to receive modification suggestions about once a week. They have ranged from dropping sand or water into the storm, or releasing huge quantities of oil on the water areas ahead of the storm, to running eight hurricane hunter planes with the wind so that the prop wash would slow down the hurricane winds.

The fantastic energy to be found in a full-blown tropical cyclone has been discussed, and it is apparent that this energy cannot be combated with force. However, the hurricane must have an Achilles heel; there must be some unstable period in its energy cycle which can be influenced with comparatively small amounts of energy—as a man can tip a delicately balanced hundred-ton boulder down a mountainside with only one hand. It is known that hurricanes build or die within a matter of hours, that they are very sensitive to influences from weather patterns outside the tropics, and that the central parts of a hurricane are in a constant state of fluctuation of such things as eye structure, wind speed, and barometric pressure. The attack point, the Achilles heel, may be the manipulation of this internal turbulence to advantage, or the answer may be in a massive assault with fleets of aircraft loaded with seeding materials. Our knowledge of hurricanes is still too meager at the present time to predict the answer to this problem.

The Navy and Weather Bureau are conducting a joint hurricane modification experiment called Project STORMFURY. Two hurricanes have been seeded so far, but no conclusive evidence has been uncovered to show that man's efforts bothered these monsters of the sea for any appreciable period. The program is going forward, and the day may come when seeding planes can be sent out to reduce the intensity of storms or to turn hurricanes away from populated areas.

CAPTAIN EDWIN T. HARDING, U. S. NAVY

In 1932, Captain Harding was graduated from the University of California at Berkeley and was appointed an Ensign in the U. S. Naval Reserve. After eight years in the school system of his native state of California—two as a high school principal—he reported for active naval service in 1941. He was an Aerological Instructor at the Pensacola Naval Air Station for a year, after which he was assigned to the Naval Postgraduate School for instruction in aerological engineering.

From the summer of 1943 until the spring of 1945, he served in various aerological capacities in the Western Pacific area, and established the U. S. Navy Weather Central at Manus in the Admiralty Islands, where he served as Executive Officer and Officer in Charge. In April 1945 he became Aerological Officer at the Naval Air Station, Lakehurst, and eight months later transferred, in a similar capacity, to the staff of Commander Western Sea Frontier, where he planned and forecasted for Operation Turtle, the record long-range Navy flight.

Since 1946, Captain Harding has served in assignments such as Officer in Charge, U. S. Navy Hurricane Weather Central, Miami, and U. S. Navy Weather Central, Guam; Officer in Charge Fleet Weather Central, Norfolk, and Project AROWA (Applied Research and Operational Weather Analysis); Aerological Officer, Staff Commander Second Fleet; Commanding Officer, Fleet Weather Central, Port Lyautey, Morocco; and Officer in Charge, Meteorology-Oceanography Curricula at the U. S. Naval Postgraduate School. In July 1962, he became Assistant Director of the Naval Weather Service in Washington, D. C.

PART II—TYPHOONS

Introduction

"The truth is, I didn't even know there was a typhoon in the area until my ship was hit by 70-knot beam winds—right after I got into San Bernardino Strait! I guess we were lucky to get out of that one." Thus spoke a captain of the Military Sea Transportation Service (MSTS).

The master of a commercial vessel en route from Guam to Yokohama related that his ship managed to work its way into the eye of an intense typhoon. Cargo damage alone was in excess of $85,000. His explanation: "I knew I could outrun that d——d typhoon, but I forgot how fast those huge swells move 'way ahead of those storms. Once those swells hit the ship, my SOA (Speed of Advance) was cut to 4.5 knots. I was a sitting duck, and the eye passed right over me!"

I recall being asked by the skipper of a Navy oiler (AO): "Say, Bill, how do the winds blow around these typhoons out here—is it clockwise or counterclockwise?" And the statement of a senior military officer was equally as startling: "Typhoons should come in pairs. Then one would destroy the other and no one would get hurt." Typhoons sometimes do "come in pairs," but the result is never one of mutual destruction. Quite the contrary.

These statements—and many similar ones which I have heard during discussions, typhoon and weather briefings, and conferences—have made it painfully obvious that a large percentage of Western Pacific (WESTPAC) military and civilian personnel who are in positions of command or authority know little about their greatest non-political peacetime threat—the typhoon.

There are probably three reasons for this:

1. The typhoon is an extremely complex phenomenon, and most treatments of the subject are highly theoretical and highly mathematical.
2. Basic information relating to typhoons is contained only in widely scattered textbooks and technical journals—not readily available.
3. Military and civilian personnel in positions of command or authority do not have an opportunity to gather all the technical meteorological material together and to decipher it.

Consequently, the purpose of this guide is to provide a readily available, readable, concise, non-mathematical, non-theoretical, well-illustrated source of operational information about one of nature's most destructive phenomena. If one ship is saved from ending up in the eye of a typhoon, or if typhoon damage at one fixed-base installation is minimized, my efforts will have been amply rewarded.

William J. Kotsch
Captain, U. S. Navy

Norfolk, Virginia
18 January 1965

Section 1 : Some Topical-Tropical Words, Phrases, and Concepts

There are many unsolved problems today with regard to tropical meteorology in general and typhoons in particular. For example, there is still no universally accepted theory on the formation of typhoons. The primary reasons for this are the lack of money and personnel for basic and applied research, and the extreme paucity of meteorological data in tropical latitudes.

It is known, however, that all motion in the earth's atmosphere results from inequalities in heating and cooling—particularly the net radiation differences between the tropical and polar regions. And since the temperature averaged over many years at all places is nearly constant, these differences must be balanced by the heat exchange between polar and tropical latitudes. The air currents which are set in motion blowing toward the poles and toward the Equator largely effect the exchange. Still unknown are the exact details of how the exchange takes place.

Nevertheless, the considerable knowledge we do have of the tropics and typhoons can be put to good operational use by ships' captains and other decision-makers in the Western Pacific (WESTPAC) area. To do this, an understanding of a few key words, phrases, and weather concepts is necessary.

Trade Winds (or Trades)

If the earth had a uniform surface, and if the planet neither rotated on its axis nor revolved around the sun, a simple atmospheric circulation would result. Since this is not the case, constantly changing large-scale movements of air take place, and a part of this air movement is called the trade winds in tropical regions.

In each hemisphere, there exists a "belt" of high pressure, centered near 30° to 35° latitude—actually several anticyclonic cells rather than a continuous belt. The winds blow from these areas of high pressure toward the Equator. Because of the earth's rotation, the moving air in the Northern Hemisphere is deflected to the right (to the west), and instead of north winds, the northeast trades result. In the Southern Hemisphere it is just the reverse. The air traveling from the high pressure area near 30° to 35° S toward the Equator is deflected to the left (toward the west). So instead of south winds, the southeast trades result.

As is to be expected, the northeast trades on approaching and crossing the Equator become northerly and northwesterly winds in the Southern Hemisphere (since they are then deflected to the left). And the southeast trades of the Southern Hemisphere on approaching and crossing the Equator become southerly and southwesterly winds in the Northern Hemisphere (since they are then deflected to the right).

The trade winds in each hemisphere (averaging between 11 and 19 knots, and slightly stronger during winter months than summer) encompass most of the tropics, and are considered to be the most constant portion of the earth's wind regime. In this regard, Figure 1 depicts a profile of "wind constancy" versus latitude. Note that the trade winds attain a constancy of 80 per cent, and remember that when it is summer in the Northern Hemisphere, it is winter in the Southern Hemisphere and vice versa.

Intertropical Convergence Zone (ICZ)

The area near the Equator where the northeast trades (of the Northern Hemisphere) and the southeast trades (of the Southern Hemisphere) come together—or converge—has been rather well established by the various prevailing-wind analyses and charts prepared by several eminent meteorologists. And the region within which the northeast and southeast trade-wind air currents flow together, or converge, is known as the Intertropical Convergence Zone, frequently abbreviated as the ICZ—or ITC in some of the literature. This region is also referred to as the Equatorial Convergence Zone, Equatorial Trough of Low Pressure, Equatorial Front, and Equatorial Shear Line by various meteorologists. However, in this guide, it will be referred to as the Intertropical Convergence Zone—the ICZ.

In general, the ICZ tends to follow the sun. It meanders northward and southward in each season, and this total annual meandering in the Western Pacific amounts to as much as 44° of latitude in extreme cases. The mean positions of the ICZ in WESTPAC for the month of February, of August, and of May and November are shown in Figures 2, 3, and 4, respectively. As seen on these charts, the ICZ reaches its northernmost latitude in August, and its southernmost latitude in February. It must be appreciated, however, that the day-to-day variations in the position of the ICZ may differ substantially from the mean position during either the winter or the summer months.

The ICZ is usually characterized by ascending air currents, considerable cloudiness, and frequent heavy showers and thundershowers. The intensity, however, varies greatly. Sometimes it is a veritable "wall of black clouds," some of whose tops extend to 55,000 feet and higher, with severe turbulence in the clouds characteristic of squall lines of higher latitudes. At other times, it is so weak and inconsequential that it is difficult to locate on the weather map, for there is very little weather or cloudiness associated with it. Its width varies from about 20 to 150 nautical miles, and as a general rule, the narrower the zone, the more intense the weather associated with it.

When the ICZ is located near the Equator, only small cyclonic circulations (vortices) develop within it. But when it migrates northward (or southward into the Southern Hemisphere), the influence of the earth's rotation becomes great enough to transfer sufficient "spin" to the converging air currents to permit tropical cyclones and typhoons to develop.

Easterly Waves (EW)

As a rule, one speaks of "waves" in a flow current when that current executes fairly sinusoidal oscillations. One of the most permanent currents in the earth's atmosphere, as mentioned above, is the tropical easterly current of the western North Pacific which, at times, oscillates in a wavelike manner. Thus, one speaks of—and deals with—waves in the tropical easterlies, or easterly waves (EW).

Easterly waves are extremely important in the WESTPAC area because of their relation to tropical-cyclone and typhoon formation. Essentially, these waves are troughs of low pressure, which are embedded in the deep easterly current located to the south of the Western Pacific high-pressure area. On the average, these

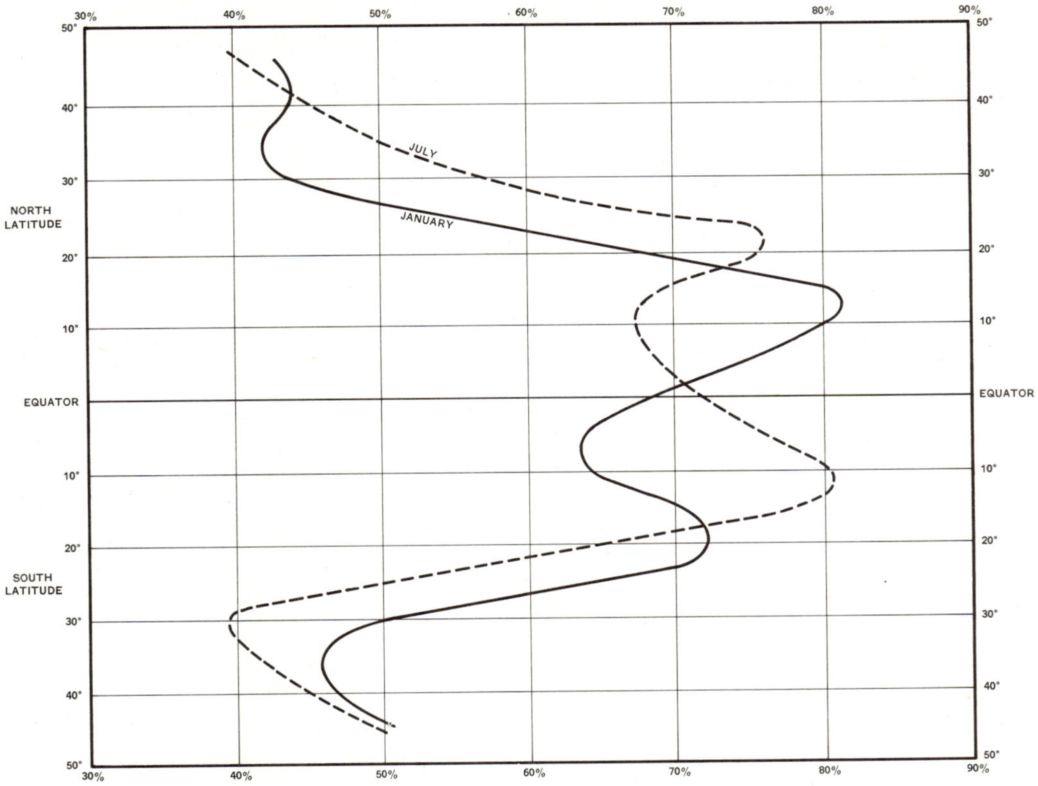

Fig. 1 *Constancy of the surface wind direction versus latitude expressed in per cent. Note that the trade winds attain a constancy of 80 per cent.*

Fig. 2 *Intertropical Convergence Zone—Mean position for February.*

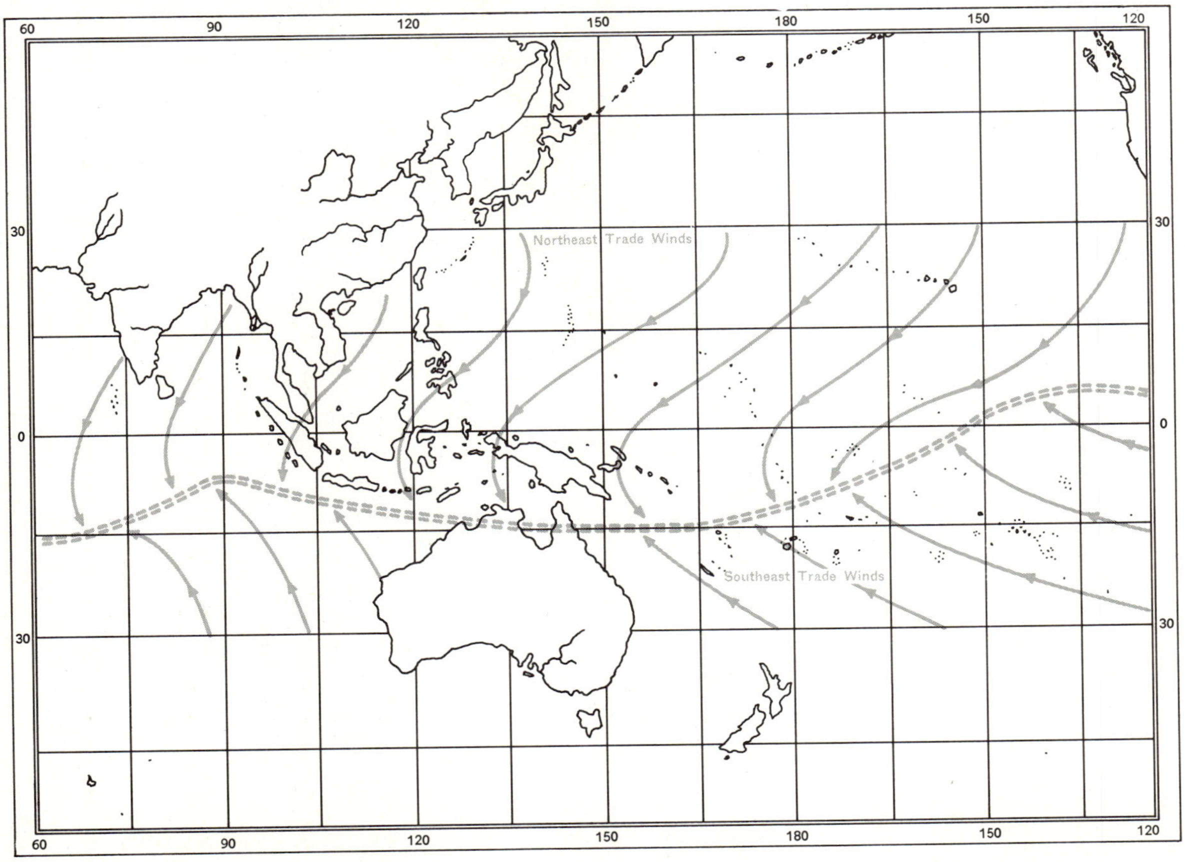

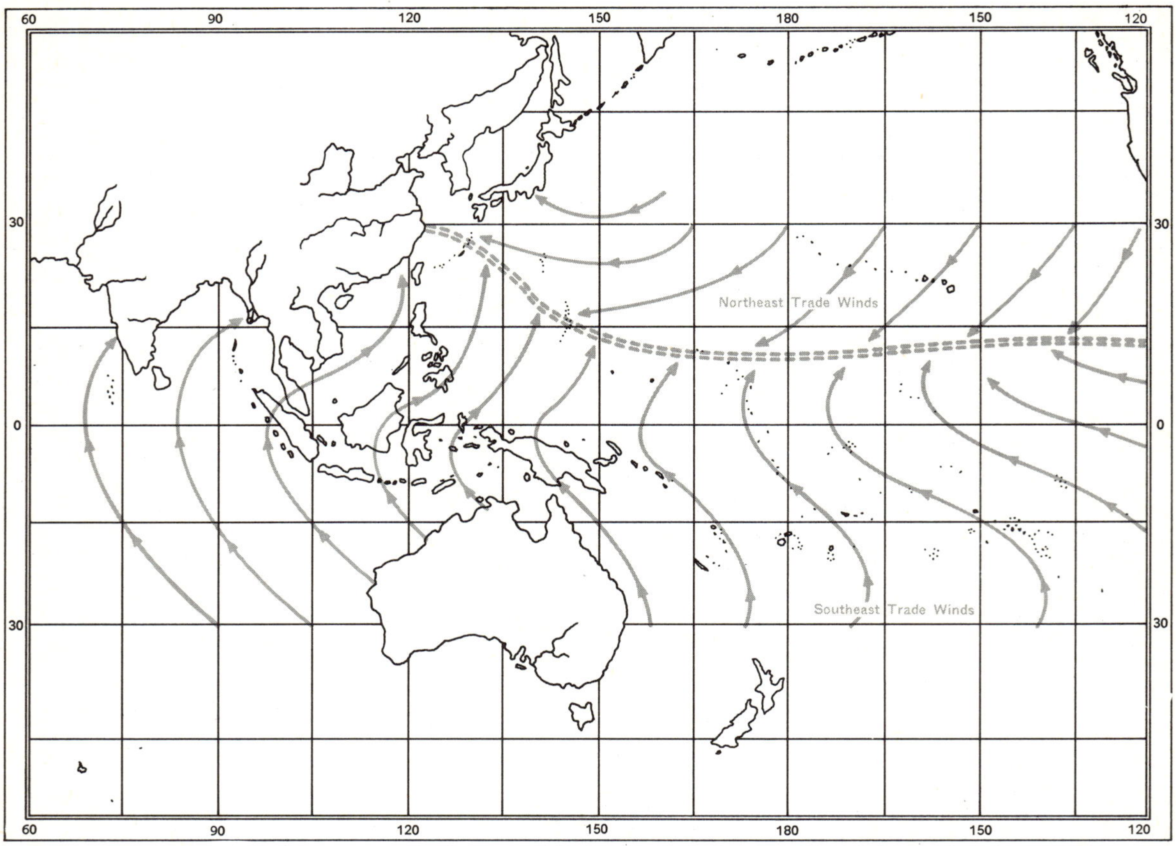

Fig. 3 Intertropical Convergence Zone—Mean position for August.

Fig. 4 Intertropical Convergence Zone—Mean position for May and November.

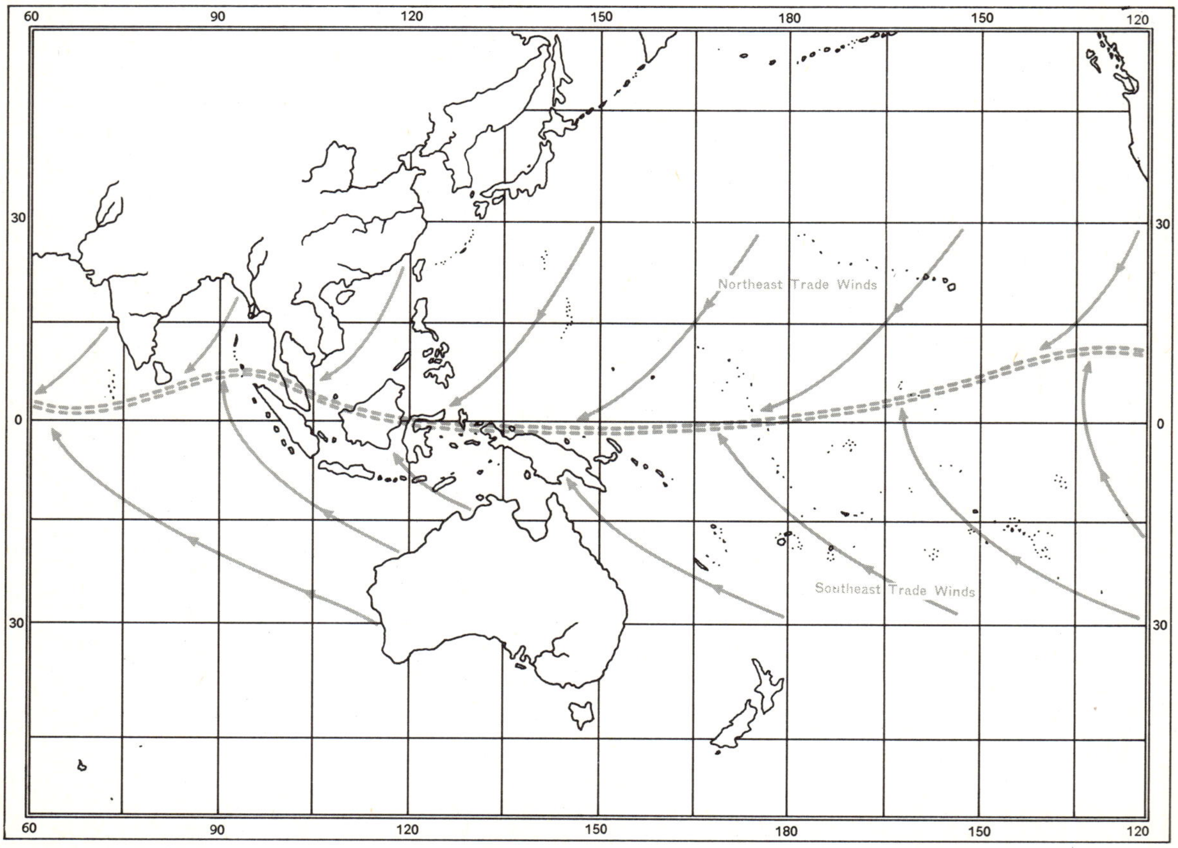

waves occur about every 15° of longitude; they have an average length of about 18° of latitude, extend vertically in the atmosphere from the earth's surface to about 26,000 feet, and travel from east to west at an average speed of 10 to 13 knots. Figure 5 illustrates the typical surface pressure pattern of easterly waves in model form. Note that the easterly waves (double lines labeled "EW") are 90° to the mean orientation of the isobars, and that the wave amplitude decreases with increasing latitude. Rather than distinct boundaries, easterly waves are zones of transition 30 to 100 miles wide in which the weather changes rather gradually, but very definitely.

To complicate things just a little, there are two types of easterly waves: stable and unstable. Although both types move from east to west and both have an average slope of about 1 to 70, a stable easterly wave slopes vertically from west to east. Consequently, the bad weather occurs behind (to the east of) a stable EW. An unstable EW slopes vertically from east to west. As a result, the bad weather occurs ahead (to the west) of the unstable EW. Vertical cross sections of stable and unstable EWs and their associated movements and weather are shown in Figures 6 and 7, respectively.

In brief summary, the weather characteristics are as follows:

	Stable EW	*Unstable EW*
West of Trough	Fair weather cumulus. Few build-ups. Strong haze. No precipitation. ENE to NE surface winds.	Heavy cumulus and thunderheads. Layers of middle and high clouds. Heavy showers, thundershowers and rain. ENE to NNE surface winds.
Close to Trough	Cumulus build-ups. High and middle clouds. Improving visibility. Scattered showers. ENE to E surface winds.	Heavy cumulus build-ups. Broken to overcast high and middle clouds. Frequent showers. ENE to E surface winds.
At Trough Line	Heavy cumulus build-ups. Broken to overcast high and middle clouds. Frequent showers. Easterly surface winds.	Cumulus build-ups. High and middle clouds. Improving visibility. Scattered showers. Easterly surface winds.
East of Trough Line	Heavy cumulus and thunderheads. Layers of middle and high clouds. Heavy showers, thundershowers and rain. ESE to SSE surface winds.	Fair weather cumulus. Few build-ups. Moderate haze. Little precipitation. ESE to SE surface winds.

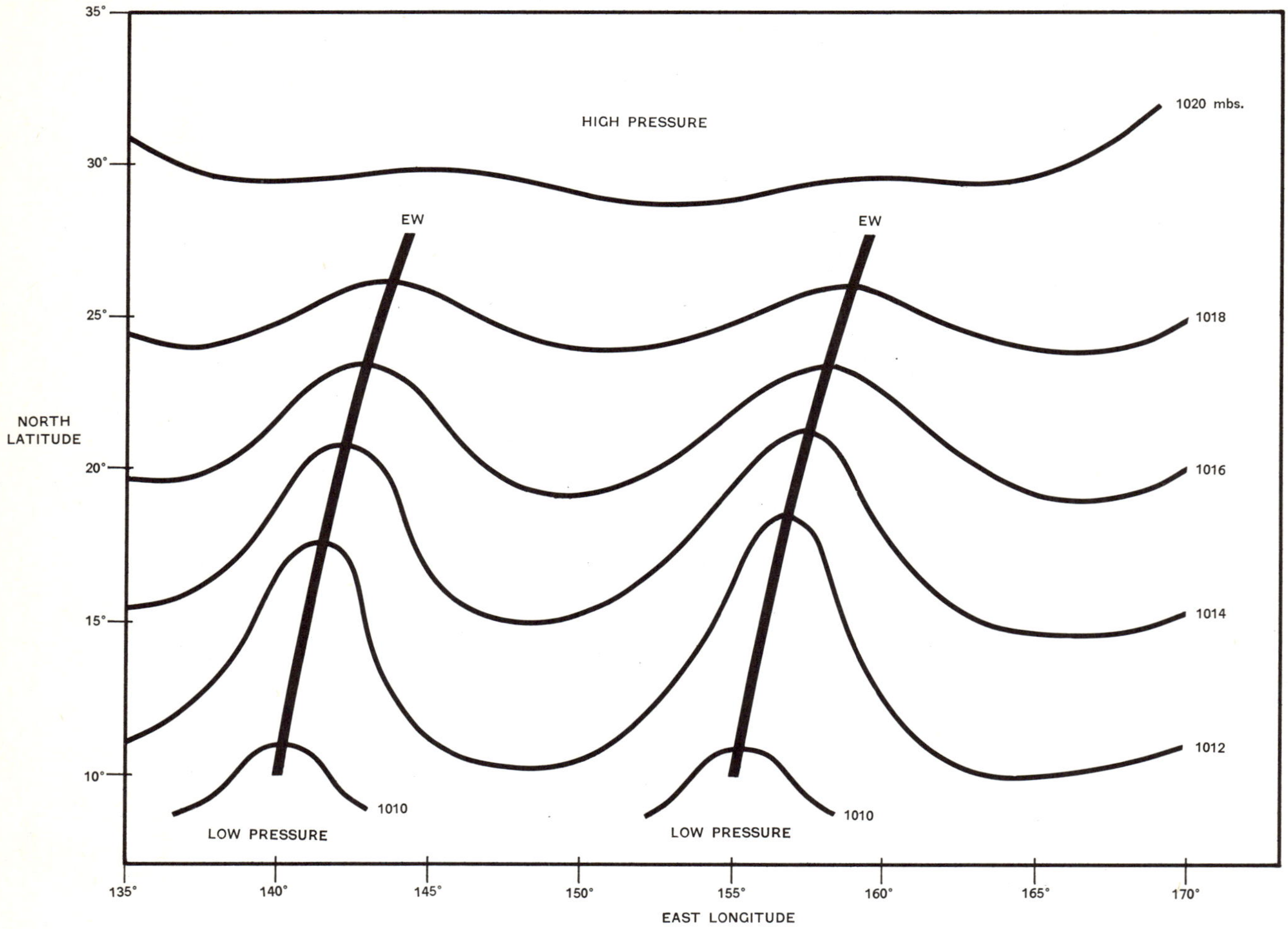

Fig. 5 *Model of easterly waves (EWs) on a surface weather chart. Note that the waves are 90° to the mean orientation of the isobars and that the wave amplitude decreases with increasing latitude.*

Fig. 6 *Vertical model of a stable easterly wave (EW), moving from east to west at 13 knots. Note that bad weather occurs behind (to the east of) the EW.*

Fig. 7 *Vertical model of an unstable easterly wave (EW), moving from east to west at 10 knots. Note that bad weather occurs ahead (to the west) of the EW.*

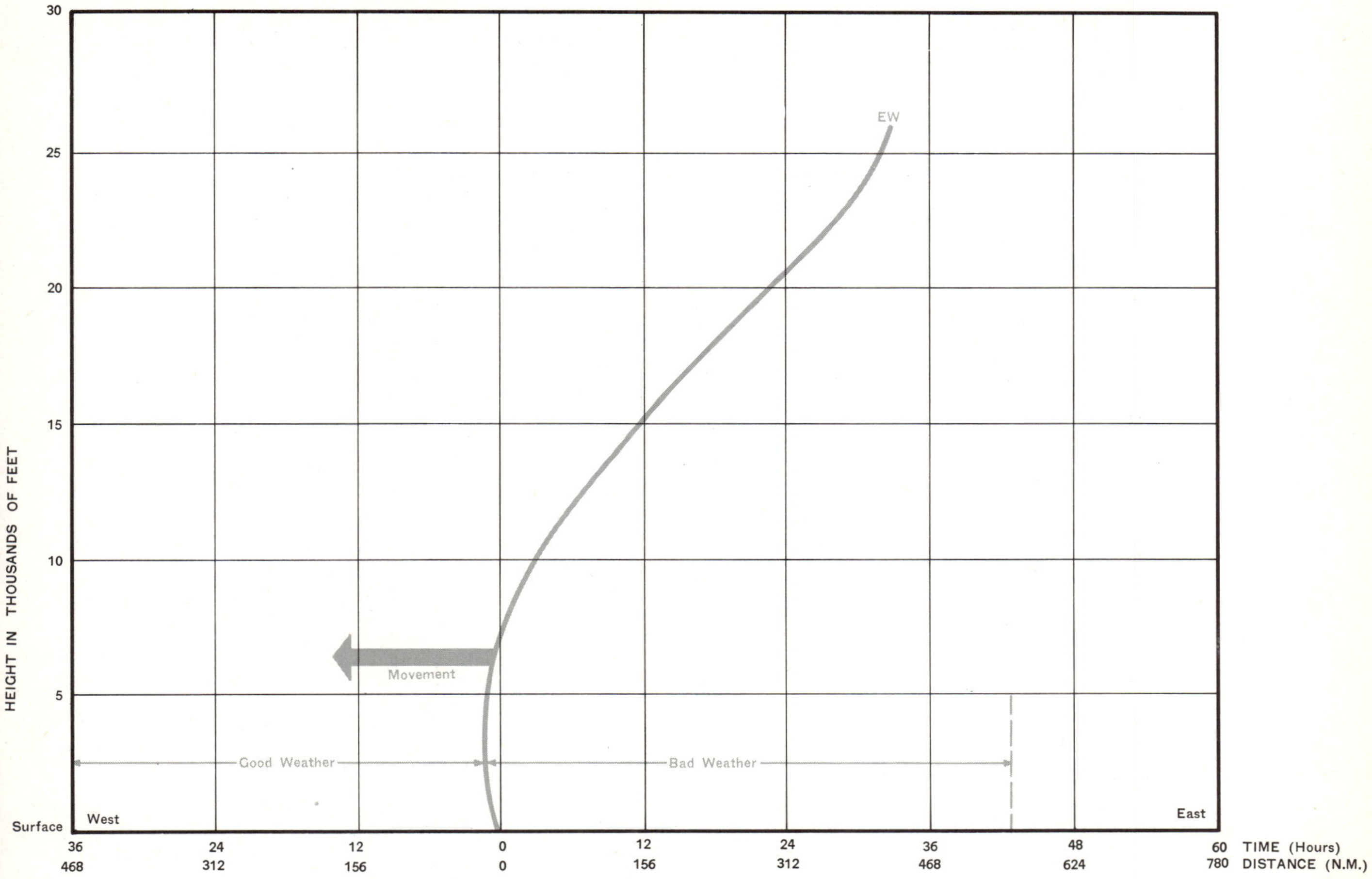
HEIGHT IN THOUSANDS OF FEET
30
25
20
15
10
5
Surface
EW
Movement
Good Weather
Bad Weather
West
East
36 24 12 0 12 24 36 48 60 TIME (Hours)
468 312 156 0 156 312 468 624 780 DISTANCE (N.M.)

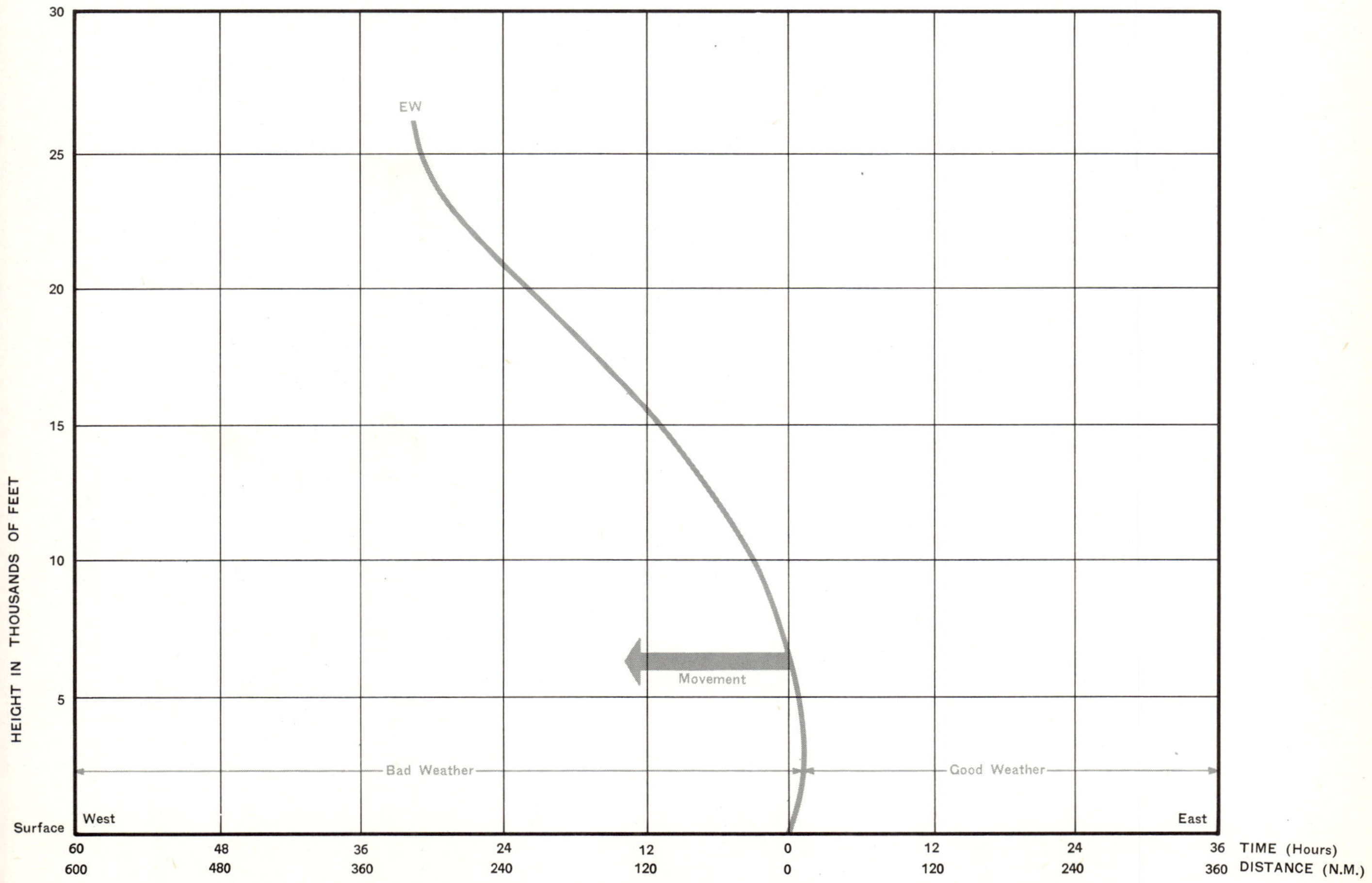
HEIGHT IN THOUSANDS OF FEET
30
25
20
15
10
5
Surface
EW
Movement
Bad Weather
Good Weather
West
East
60 48 36 24 12 0 12 24 36 TIME (Hours)
600 480 360 240 120 0 120 240 360 DISTANCE (N.M.)

Section 2 : What Is a Typhoon?

Before this question—What is a typhoon?—can be answered properly, it is necessary to backtrack slightly and define the word "cyclone." Stated simply, a cyclone is an atmospheric pressure system in which the barometric pressure decreases progressively to a minimum value at the center, and toward which the winds blow spirally inward from all sides, resulting in a "lifting" of the air and eventually in clouds and rain. In the Northern Hemisphere, the winds spiral in toward the center in a counterclockwise direction. In the Southern Hemisphere, the winds spiral inward in a clockwise direction.

Cyclones are divided into two basic groups, tropical (warm-core) and extratropical (cold-core), depending upon the area of origin and the characteristics of the surrounding air masses. The typhoon is, of course, a tropical cyclone, and is undoubtedly the most "ideal" (and destructive) vortex in the earth's atmosphere.

By joint and international agreement, tropical cyclones in the WESTPAC area are classified according to their intensity, as follows:

Tropical Depression
Maximum sustained winds not exceeding 33 knots.

Tropical Storm
Maximum sustained winds between 34 and 63 knots.

Typhoon
Maximum sustained winds 64 knots or higher.

At full maturity, the typhoon is a vast whirlwind of fantastic violence. It is nature's second most intense and feared phenomenon, exceeded only by the tornado in violence. Maximum sustained winds of 180 knots, with gusts exceeding 225 knots, are not at all uncommon in WESTPAC's typhoons. Typhoon Karen, which devastated the entire island of Guam the night of 11-12 November 1962, packed 150-knot sustained winds with gusts well over 185 knots.

In total damage and loss of human life, typhoons by far exceed any other natural catastrophe. They have caused the sinking of many ships at sea and in port. Many areas have been inundated with great loss of life (southern Thailand in October 1962). Typhoons have played important roles many times, such as in 1944-45 during World War II, in November 1952 at the time of the first H-bomb "shot," and in October 1962 during Astronaut Schirra's orbital flight. To be in a typhoon is a terrifying experience which remains vivid in one's memory for many years. Possibly for this reason they have inspired many stories.

In various parts of the world, typhoons are known by different names, as follows:

Area	*Name*
WESTPAC	Typhoon
Philippines	Baguio or Baruio
Indian Ocean	Cyclone
Australia	Willy-Willy
North Atlantic	Hurricane
Caribbean Sea	Hurricane
Gulf of Mexico	Hurricane
Mexico	Cordonazo
Haiti	Taino

The word "typhoon" is derived either from the Cantonese *t'ai fung* (a great wind), from the Arabic *tufan* (smoke), or from the Greek *typhon* (a monster). And even Aristotle used the word *typhon* for a wind-containing cloud.

To answer the question posed at the beginning of this section, a typhoon is a severe tropical cyclone in the WESTPAC area with maximum sustained winds of 64 knots or higher—and a good thing to stay well clear of, if at all possible.

Formation

As can be seen in Figure 8, the "breeding ground" of WESTPAC's typhoons extends from near the Equator to 25° N, and from 110° E to 180° E—a tremendous area of approximately 6,300,000 square miles, and an area which one can hardly avoid during a normal WESTPAC tour or deployment.

During the 10-year period 1952-61, inclusive, there were 184 typhoons; the triangles in Figure 8 show the locations in which they began to form. The last two digits of each 4-digit group indicate the calendar year in which that particular typhoon formed, and the first two digits indicate the typhoon's sequence number for that year. Thus, 0452 means "the 4th typhoon of 1952," 2061 means "the 20th typhoon of 1961," etc. Note that approximately 34 per cent of WESTPAC's typhoons formed within 500 nautical miles of the island of Guam. Figure 9 shows where these tropical disturbances first reached typhoon intensity (maximum sustained winds 64 knots or higher).

Although many scientific papers have been published on the subject of typhoon formation, there is still no universally accepted theory. There are the convectional hypothesis, the frontal hypothesis, the waves in a baroclinic easterly current theory, the dynamic instability theory, and several others. However, each theory deals with only one or two specific aspects of typhoon formation. There is no composite treatment of all aspects available at the present time. This is not too important, though, for the purpose of this guide, which is intended for operators and decision-makers, not theorists.

Most of the theoretical and operational meteorologists agree quite well on one aspect—the surface sea temperature must exceed 26° Centigrade (78.8° Fahrenheit) to enable the formation of a tropical cyclone or typhoon. Dr. A. Kasahara of Japan did considerable work along these lines in 1954, and Figure 10 is a schematic chart of the results he obtained. This chart shows the average surface sea temperatures in WESTPAC during the summer months, as well as the number of tropical cyclones in each 5° × 5° latitude/longitude square during the 10-year period 1940-49, inclusive.

The following is a list of conditions favorable for typhoon formation and observable by, or available to, the non-meteorologist:

1. Surface sea temperature higher than 26° C. (78.8° F.)
2. Below-normal pressure in low latitudes (less than 1004 millibars, or 29.65 inches), and above-normal pressure in high latitudes.
3. An existing tropical disturbance of some sort at the earth's surface.
4. Movement of the disturbance at a speed less than 13 knots.
5. Easterly winds decreasing in speed with height, but extending to at least 30,000 feet.
6. Heavy rain or rain showers in the area.

It is the writer's belief that most typhoons in the WESTPAC area develop from a junction vortex or an embedded vortex, as shown in Figure 11. When two vortices exist, it is frequently very difficult for the typhoon forecasters to predict well in advance which vortex will eventually develop to full typhoon intensity. And quite often the wind circulation around one vortex will "mask" the existence of the second vortex. Great care must be exercised in detailed meteorological analyses, which are made possible only through aircraft reconnaissance; otherwise, completely erroneous information relative to the tropical disturbance's direction and speed of movement will result.

Frequency

In the WESTPAC area, there is no so-called "typhoon season." Typhoons can—and do—occur in every month of the year, as Figure 12 will substantiate. This chart shows the number of typhoons which occurred in each month during the 11-year period 1952-62, inclusive. Atlantic sailors are more fortunate in this regard, for they "enjoy" a true hurricane season, from late May until early November. And in the calendar year 1962, the Atlantic was subjected to only three hurricanes, while WESTPAC suffered the brunt of 24 full-blown typhoons.

Although no true typhoon season exists, it is obvious from Figure 12 that 90 per cent of the typhoons in WESTPAC occurred between June and late December, during this 11-year period, with a maximum of 22.6 per cent in August and a minimum of 0.6 per cent in February. All of these figures and calculations include only full-fledged typhoons with maximum sustained winds 64 knots or higher. Tropical storms with maximum winds of 63 knots or less have not been included, although they themselves constitute quite a blow.

Gordon E. Dunn (Director, National Hurricane Center, Miami) and Banner I. Miller (Research Meteorologist, National Hurricane Center) state in their book *Atlantic Hurricanes:* "During recent years there has been frequent speculation on the possibility that the annual number of hurricanes is on the increase. . . . This increase is believed to be due in part to better detection, but there also appears to have been a significant increase in the number of hurricanes beginning around 1930. A gradual warming of the atmosphere apparently began about that time, and the greater number of tropical cyclones is probably due to that warming trend." More recently, the warming trend has apparently been reversed, for there were fewer Atlantic tropical cyclones in the years 1956 and 1957.

This is not quite true with WESTPAC's typhoons, as Figure 12 will clearly show. The years of minimal below-average numbers of typhoons were 1953, 1954, and 1959. And since the year 1959, as seen in Figure 13, there has been an "unhealthy," troublesome, and definite upward trend in the total number of typhoons per annum. The reasons for this are not precisely known at this time, and the entire subject merits detailed study and research.

Tracks and Movement

Forecasting the future track or movement of a typhoon is one of the most challenging and difficult—and sometimes one of the most gratifying—tasks that any operational meteorologist has to face. There is so much at stake.

Today—as the result primarily of aircraft reconnaissance, land radar reports, and the accumulation of meteorological data through the years—we know that "simple solutions" to typhoon movement almost never provide the correct answer. So-called "abnormal" tracks occur far too often. Too many typhoon tracks exhibit humps, loops, southward displacement

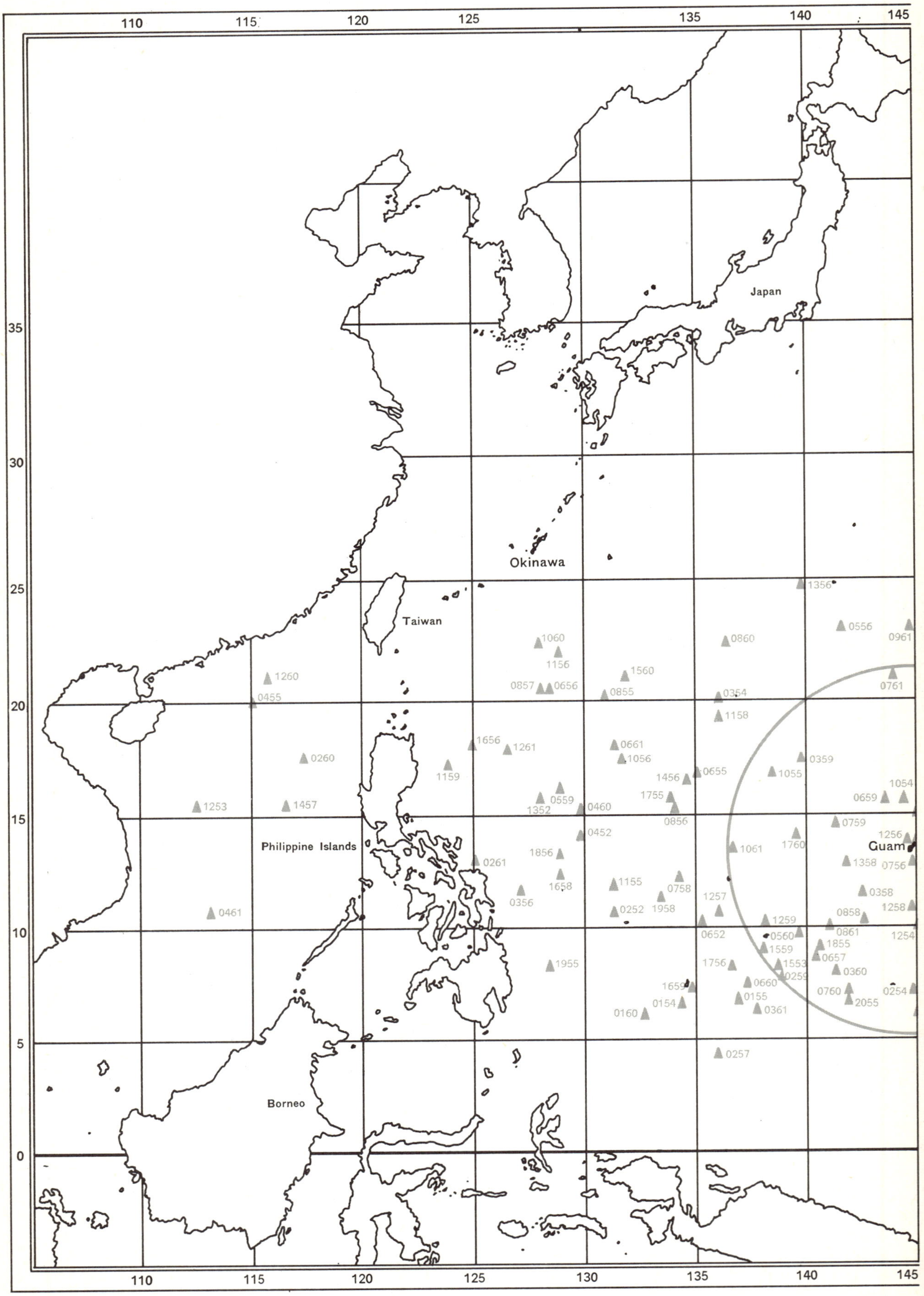

Fig. 8 *The "breeding ground" of WESTPAC's typhoons—from near the Equator to 25° N, and from 110° E to 180° E. Triangles show where the 184 typhoons between 1952-61, inclusive, began to form. Last two digits indicate calendar year typhoon formed. First two digits indicate the typhoon's sequence number for that year. 0452 means "4th typhoon in 1952," 2061 means "20th typhoon in 1961," etc. Note that 34 per cent of typhoons begin to form within 500 N.M. of Guam.*

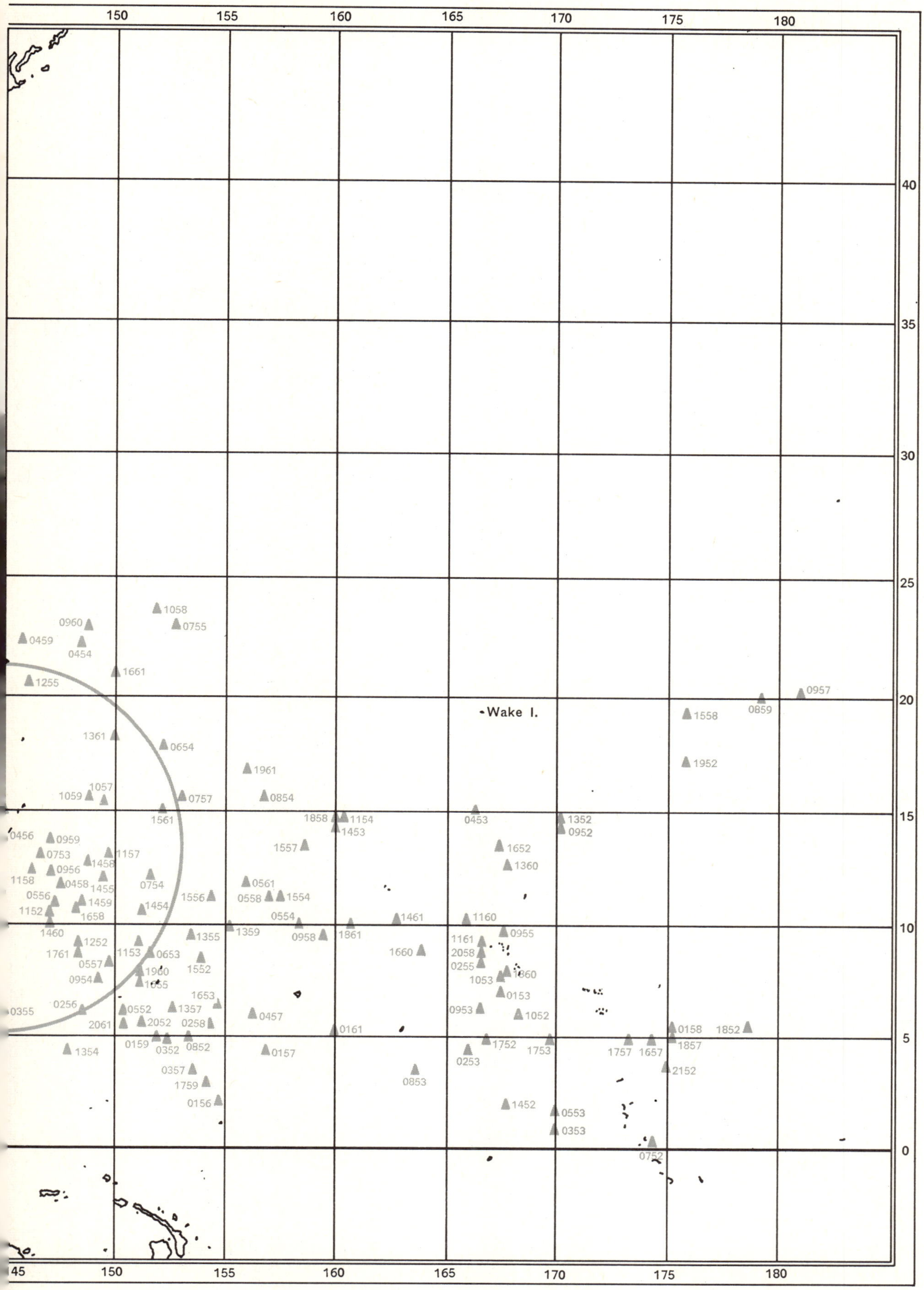
150
155
160
165
170
175
180
40
35
30
25
20
15
10
5
0
Wake I.
1058
0755
0960
0459
0454
1661
1255
0957
0859
1558
1361
0654
1961
1952
1057
1059
0757
0854
1561
1858
1154
1453
0453
1352
0952
0456
0959
0753
1458
1157
1557
1652
1158
0956
0458
1455
0754
0561
1360
0556
1459
1556
0558
1554
1152
1658
1454
0554
1461
1160
1460
1252
1355
1359
0958
1861
0955
1761
1153
0653
1552
1161
1660
2058
0557
1960
0255
1860
0954
1055
1053
1653
0153
0355
0256
0552
1357
0457
0953
1052
2061
2052
0258
0161
0158
1852
1752
1354
0159
0352
0852
0157
0253
1753
1757
1657
1857
0357
0853
2152
1759
0156
1452
0553
0353
0752
45
150
155
160
165
170
175
180

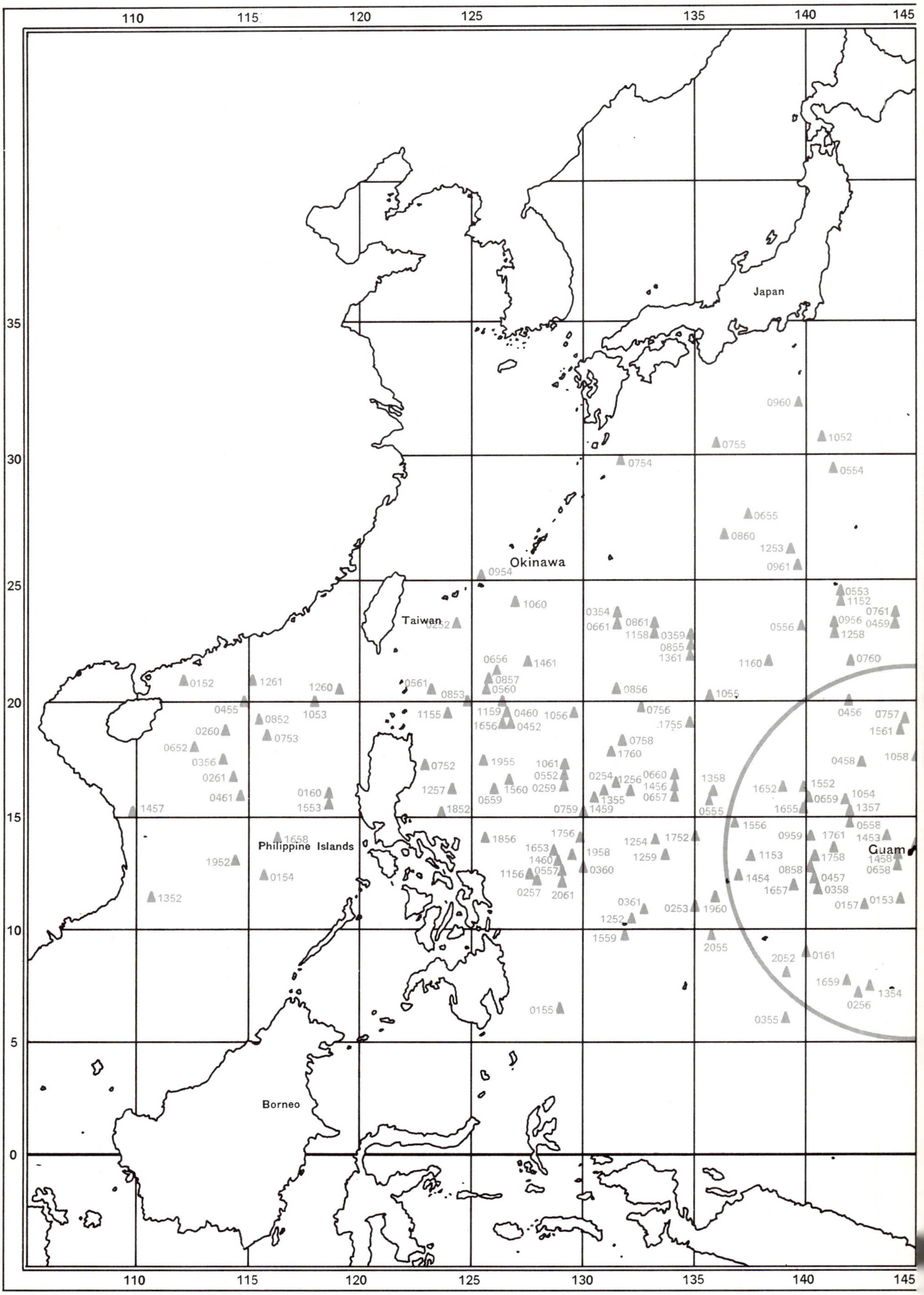

Fig. 9 *Locations in WESTPAC where the 184 tropical disturbances between 1952-61, inclusive, first reached typhoon intensity (maximum sustained winds 64 knots or higher). Note the slight shift in the triangles to the North and West from those of Fig. 8. The 4-digit groups have the same connotation as in the previous figure.*

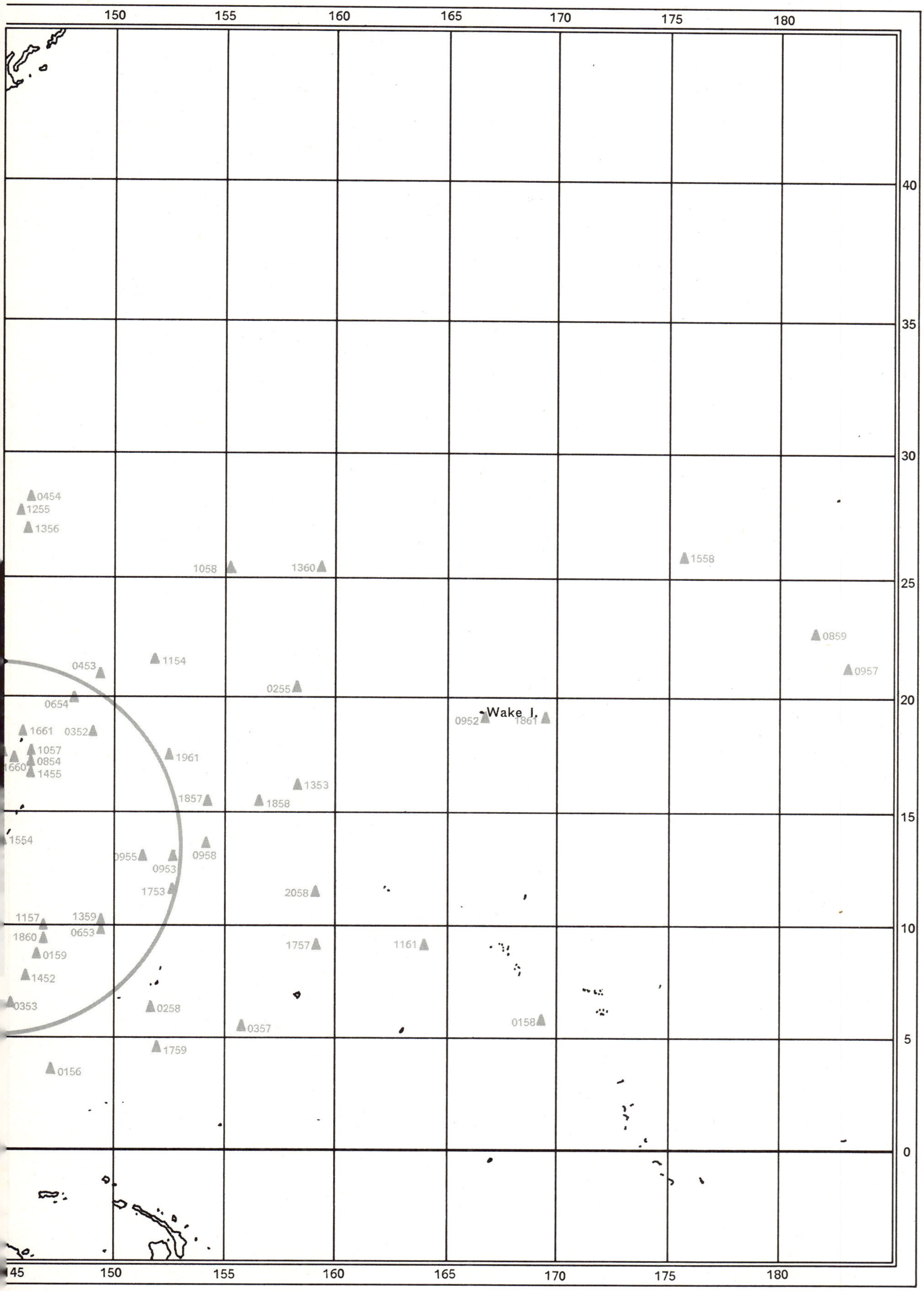

150
155
160
165
170
175
180
40
35
30
25
20
15
10
5
0
0454
1255
1356
1058
1360
1558
0859
0957
0453
1154
0255
0654
1661
0352
1057
0854
1660
1455
1961
Wake I.
0952
1861
1353
1857
1858
1554
0955
0953
0958
1753
2058
1157
1359
1860
0653
0159
1757
1161
1452
0353
0258
0357
0158
1759
0156
145

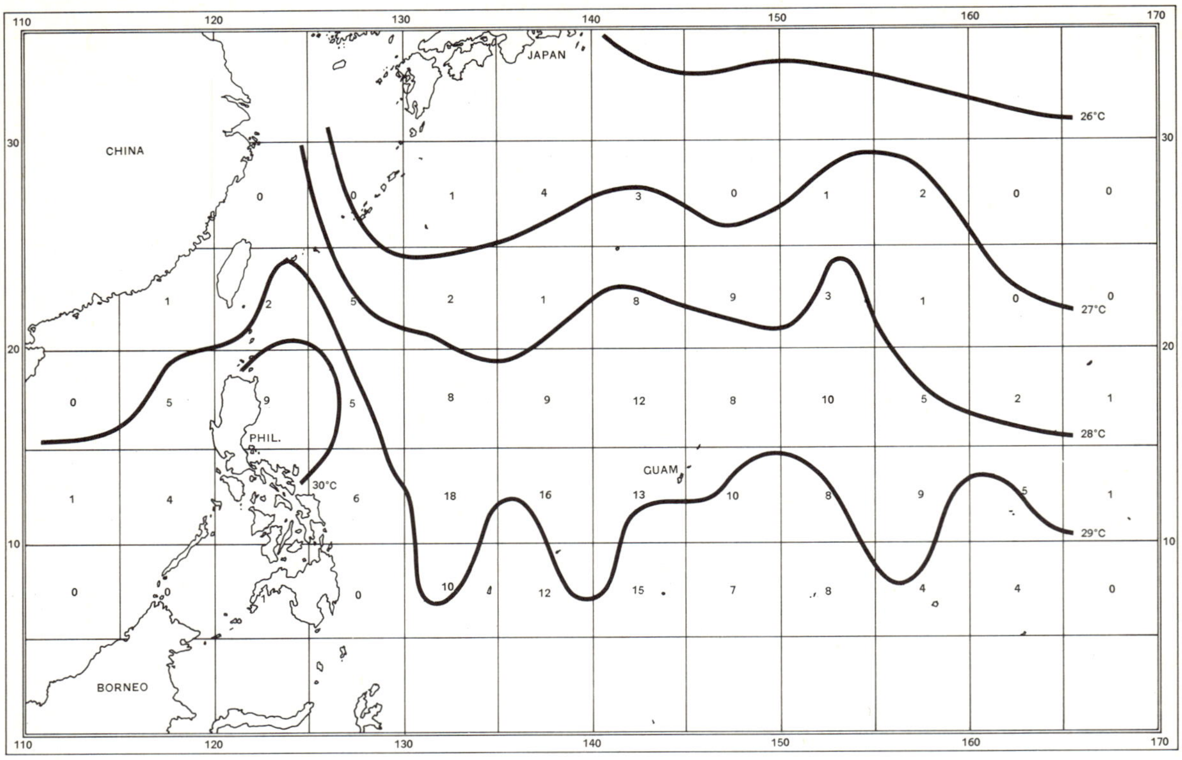

Fig. 10 *Schematic chart showing average surface sea temperatures in WESTPAC during summer months, and numbers of tropical cyclones in each 5° × 5° lat./long. square during the 10-year period 1940-49, inclusive. Note that tropical cyclones and typhoons form where surface sea temperature exceeds 26° C. (78.8° F.) (After Dr. A. Kasahara, 1954.)*

Fig. 11 *When both a junction vortex and an embedded vortex exist, it is frequently difficult to predict well in advance which vortex will develop to full typhoon intensity. Sometimes one vortex "masks" the existence of the other, leading to erroneous forecasts of speed and direction of movement. (After Lieutenant Colonel L. H. Hutchinson, USAF, 1956.)*

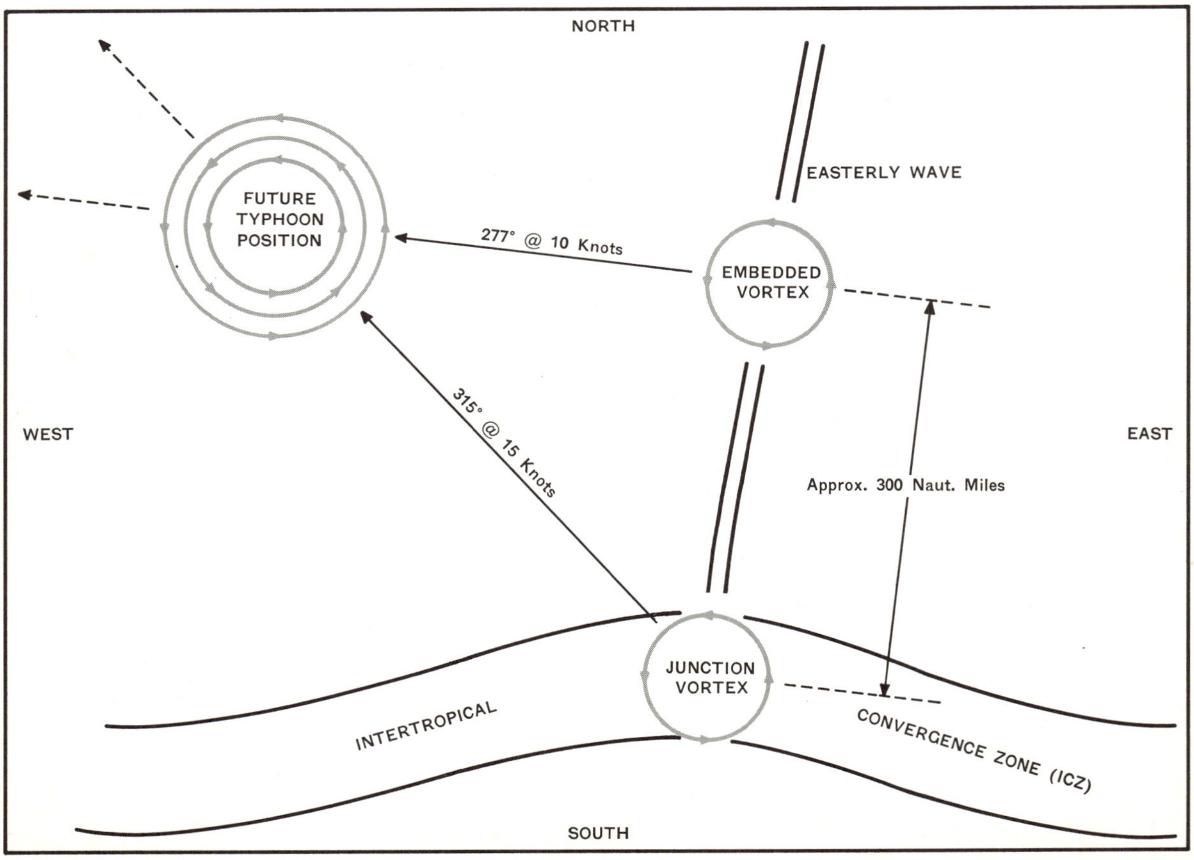

Typhoons	Jan.	Feb.	Mar.	Apr.	May	Jun.	Jul.	Aug.	Sep.	Oct.	Nov.	Dec.	Total
1952	—	—	—	—	—	3	1	3	3	5	3	3	21
1953	—	1	—	—	1	1	1	5	2	4	1	1	17
1954	—	—	—	—	1	—	1	4	4	2	3	—	15
1955	1	—	1	1	—	1	5	3	3	2	1	1	19
1956	—	—	1	1	—	—	2	4	5	1	3	1	18
1957	1	—	—	1	1	1	1	2	5	3	3	—	18
1958	1	—	—	—	1	2	5	3	3	3	1	1	20
1959	—	—	—	1	—	—	1	5	3	3	2	2	17
1960	—	—	—	1	—	2	2	8	—	4	1	1	19
1961	—	—	1	—	2	1	3	3	5	3	1	1	20
1962	—	—	—	1	2	—	5	7	2	4	3	—	24
11-Year Average	0.27	0.09	0.27	0.55	0.73	1.00	2.45	4.27	3.18	3.09	2.00	1.00	18.92
Frequency	3	1	3	6	8	11	27	47	35	34	22	11	208
Per cent Frequency	1.4	0.6	1.4	2.9	3.8	5.3	13:0	22.6	16.8	16.3	10.6	5.3	100

Fig. 12 ***Frequency of typhoons in WESTPAC. Note that there is no typhoon season; typhoons have occurred in every month. However, 90 per cent of typhoons occur between early June and late December. Maximum of 22.6 per cent in August and minimum of 0.6 per cent in February.***

Fig. 13 ***Typhoon trend in WESTPAC during the 12-year period 1952-63, inclusive.***

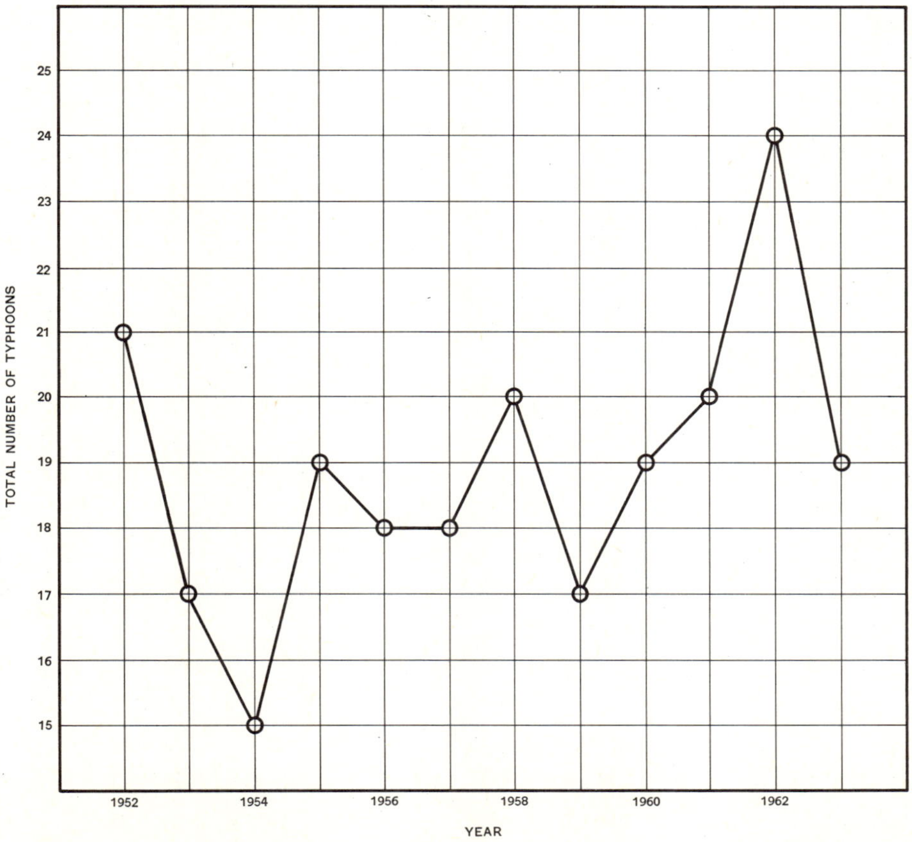

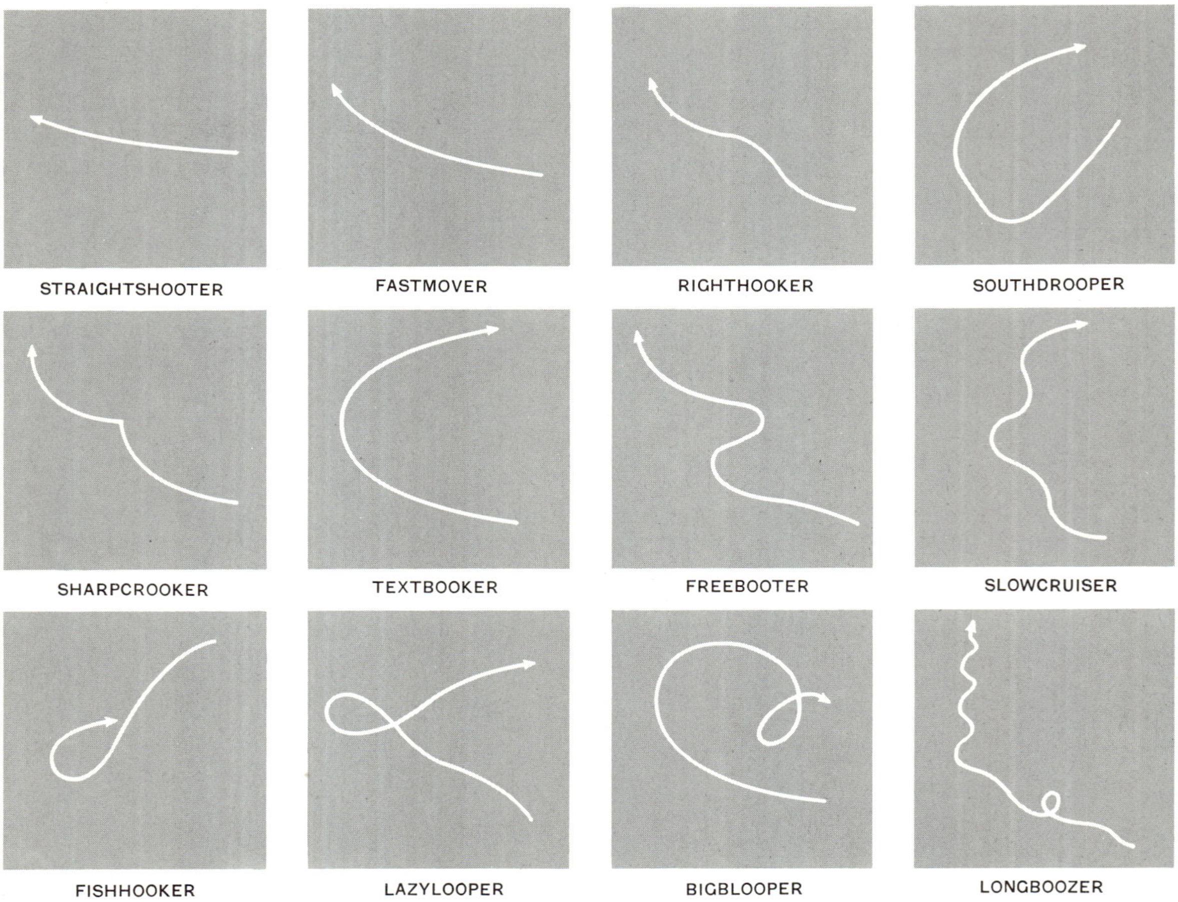

Fig. 14 Basic typhoon tracks in WESTPAC.

accompanied by rapid intensification, a staggering motion, abrupt changes in direction and/or speed of movement, etc.

No two recorded typhoon tracks have ever been exactly the same. Yet, an examination of many hundreds of typhoon tracks reveals that there are 12 basic tracks of typhoons in WESTPAC. These are depicted in Figure 14; they have been appropriately labeled by the writer, ranging from "straightshooter" to "longboozer." It must be remembered, however, that many minor variations to these basic tracks occur.

Typhoons tend to move under the influence of both internal and external forces. Internal forces arise within the circulation of the typhoon itself. External forces are applied by the currents which surround the typhoon on all sides and carry it along.

Despite all of the irregularities illustrated in Figure 14, almost all typhoons have one thing in common: a tendency to move eventually toward higher latitude. This tendency, shared by cyclones at all latitudes, suggests propulsion by an internal force. Additionally, it can be demonstrated mathematically that, in general, typhoons will move in the direction of the "steering current" in which they are embedded and with a certain percentage of its speed. This steering current is defined as the pressure-weighted mean flow from the earth's surface upward to 300 millibars (roughly 30,000 feet) over a band 8 degrees latitude in width and centered over the typhoon.

In order to calculate and forecast the future movement and tracks of typhoons, surface and upper-air weather data are essential. Unfortunately, in the WESTPAC area, the normal situation is one weather report per 226,000 square miles—leaving a great deal to be desired.

In WESTPAC, the term "recurvature" in typhoon work refers to the change in direction of movement from a westerly to an easterly component. That is, a typhoon "curves" from a course of 275° to 350°, but it "recurves" from a course of 350° to 025°. And recurvature is one of the primary problems in typhoon-track prediction.

Prior to recurvature, the small-scale track of a typhoon is usually sinusoidal, as shown in Figure 15, and the speed of movement is normally relatively slow (4 to 13 knots). After recurvature, into higher latitudes, on a NNE to ENE track, the typhoon's movement is usually straight-line and speeds of movement of 40 to 50 knots are not uncommon.

Figures 16 through 29 depict all typhoon tracks in WESTPAC for the 10-year period 1950-59, inclusive. These should be carefully scrutinized by all captains and decision-makers in the Western Pacific—and those to be deployed to this area. »»→

Fig. 15 ***Prior to recurvature—while the typhoon is on a westerly heading—the small-scale track is usually sinusoidal, and the speed of movement is normally between 4 to 13 knots. After recurvature into higher latitudes—when the typhoon is on an easterly heading—the movement of the typhoon is usually straight-line at accelerated speeds ranging from 20 to 50 knots.***

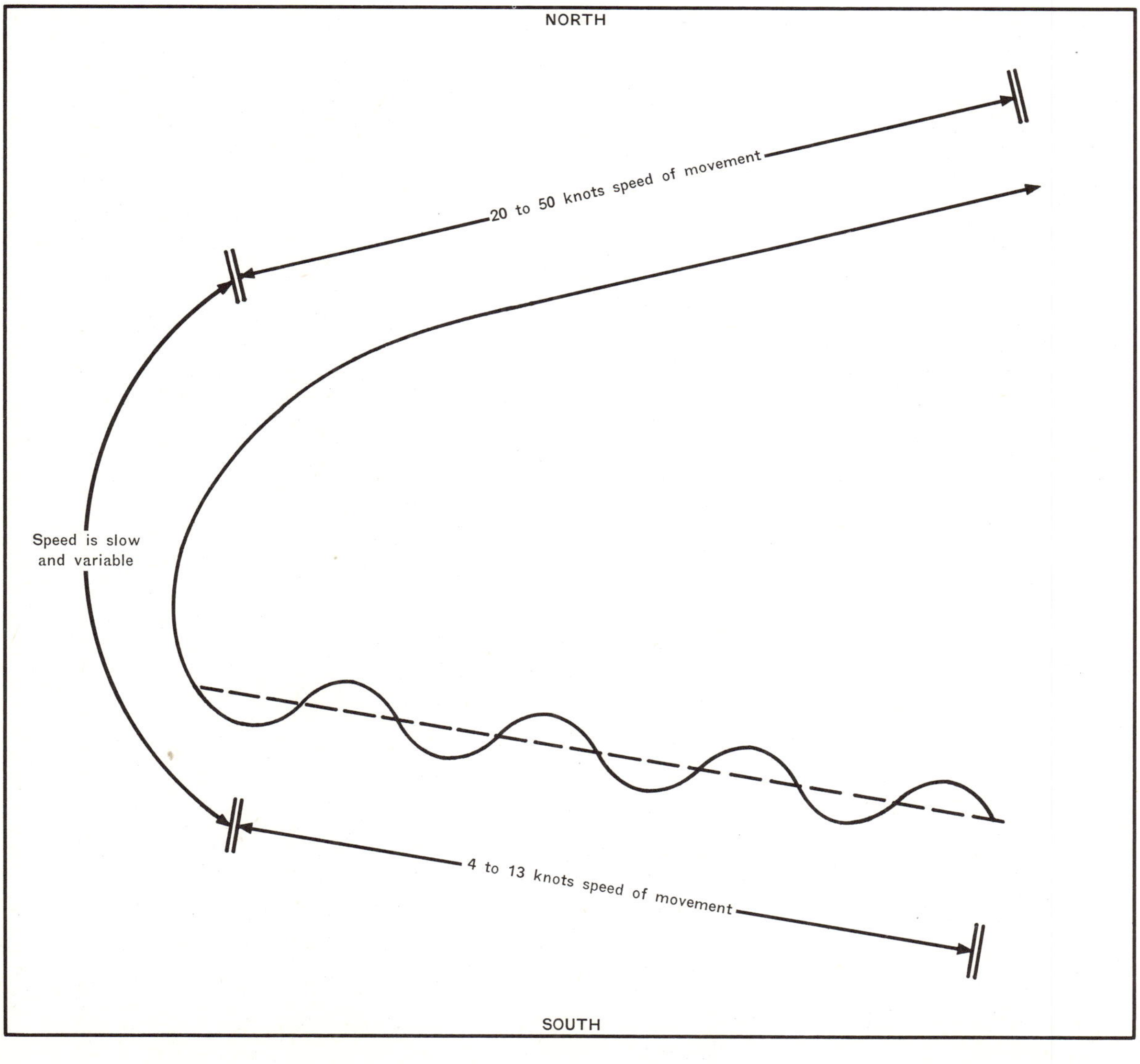

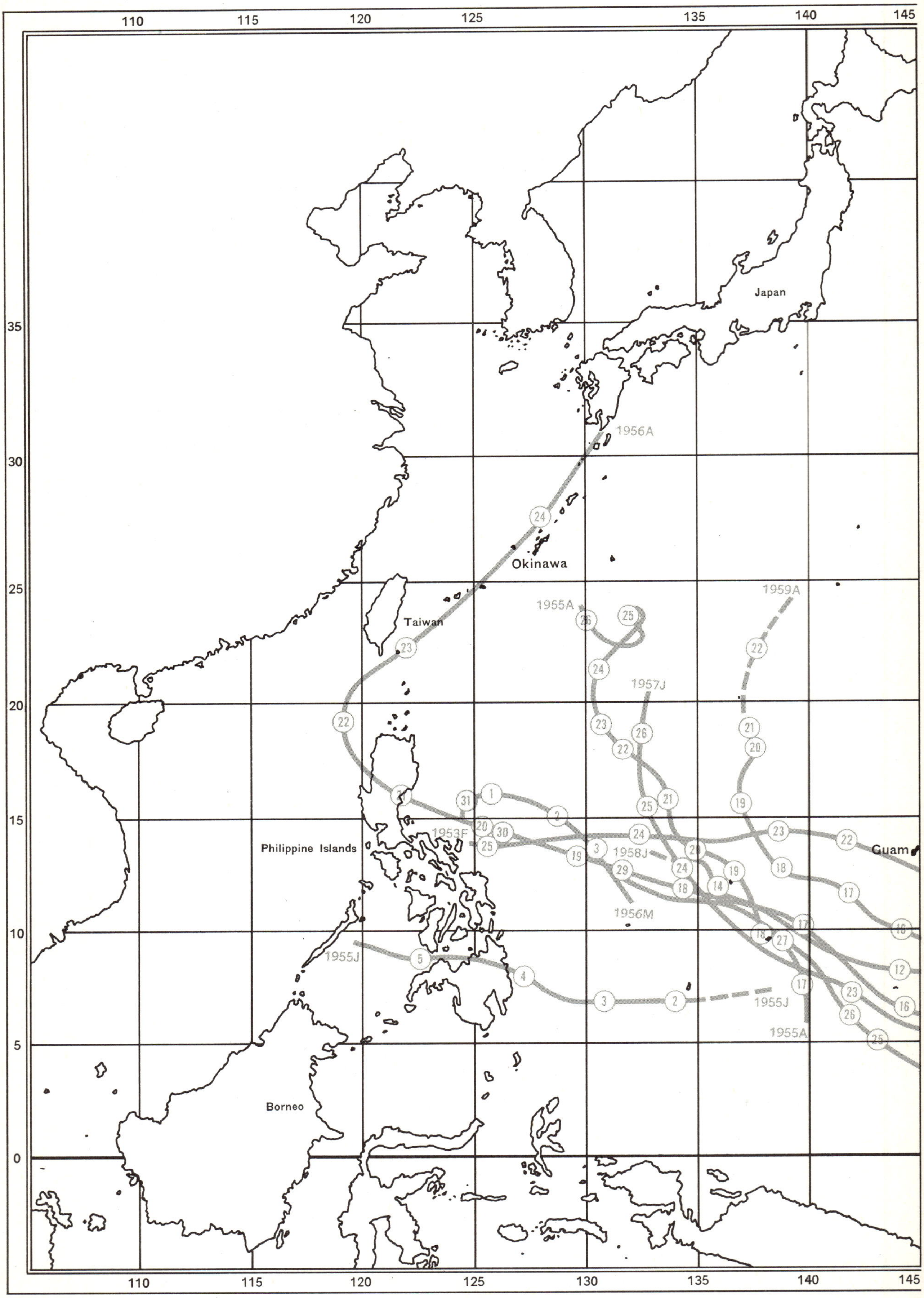

Fig. 16 Typhoon Tracks—January-February-March-April 1950-59. (Circled numbers indicate dates of each month.)

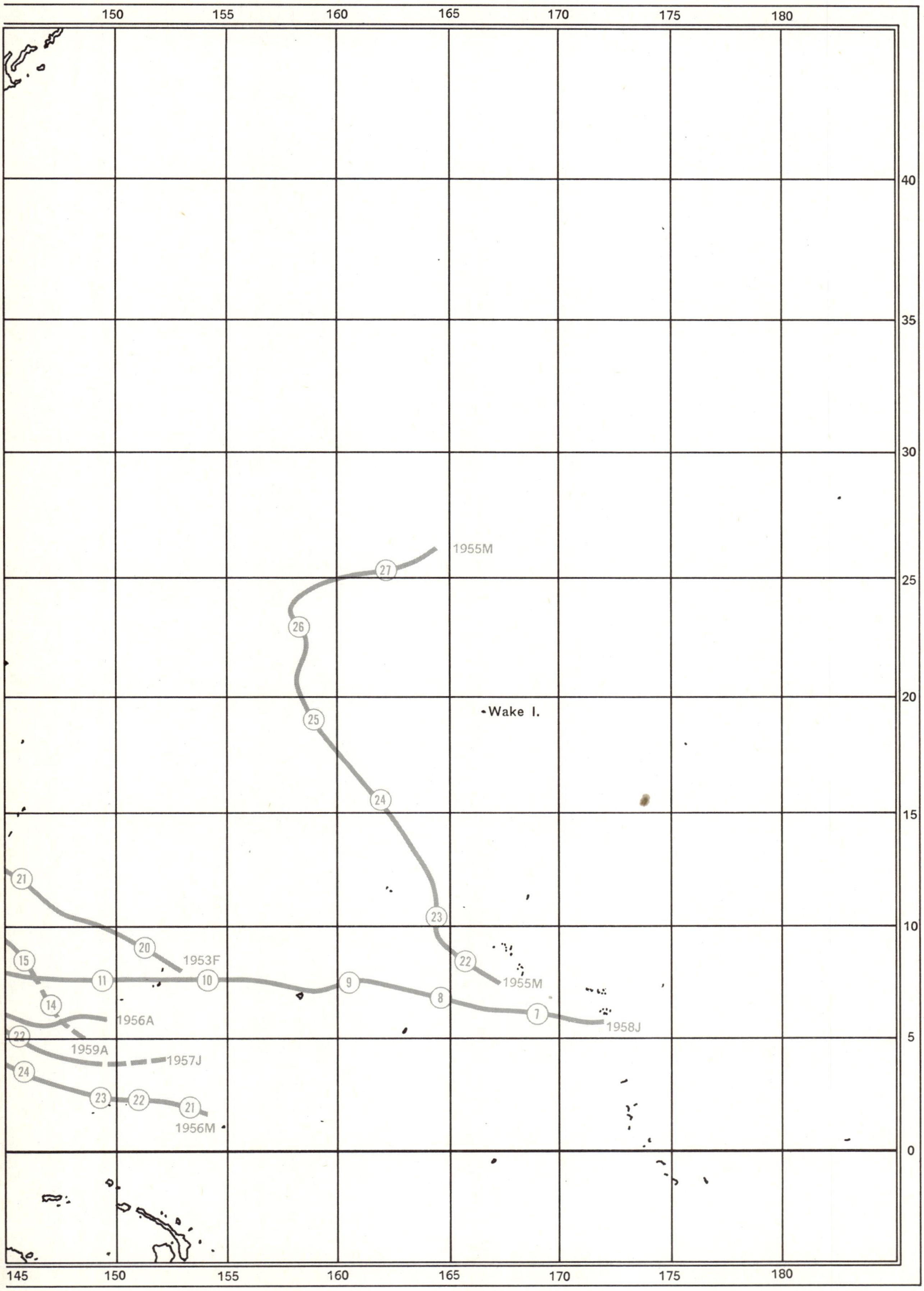
150
155
160
165
170
175
180
40
35
30
25
20
15
10
5
0
145
1955M
Wake I.
1953F
1956A
1959A
1957J
1956M
1958J
27
26
25
24
23
22
21
20
15
14
11
10
9
8
7

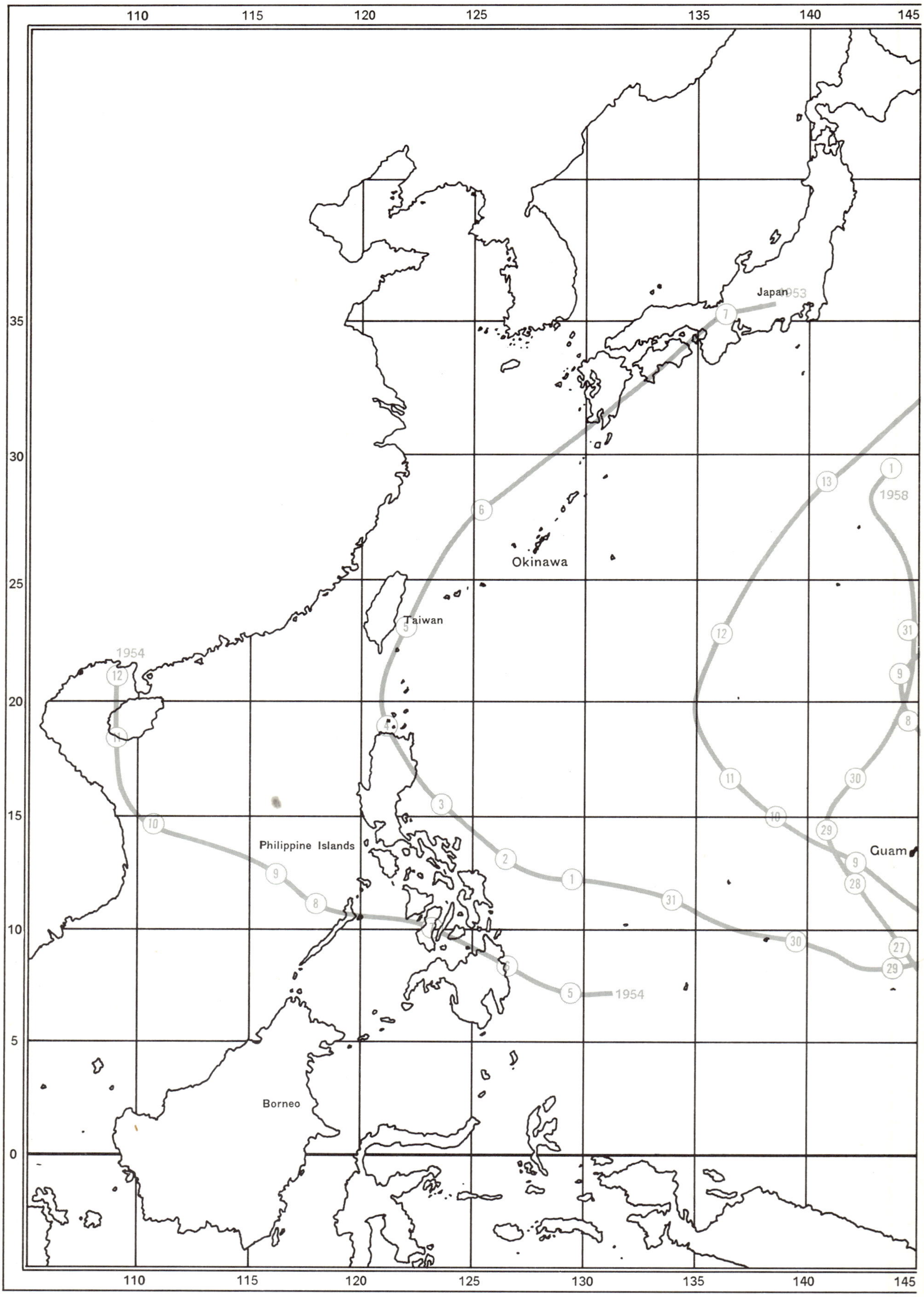

Fig. 17 Typhoon Tracks—May 1950-59. (Circled numbers indicate dates of the month.)

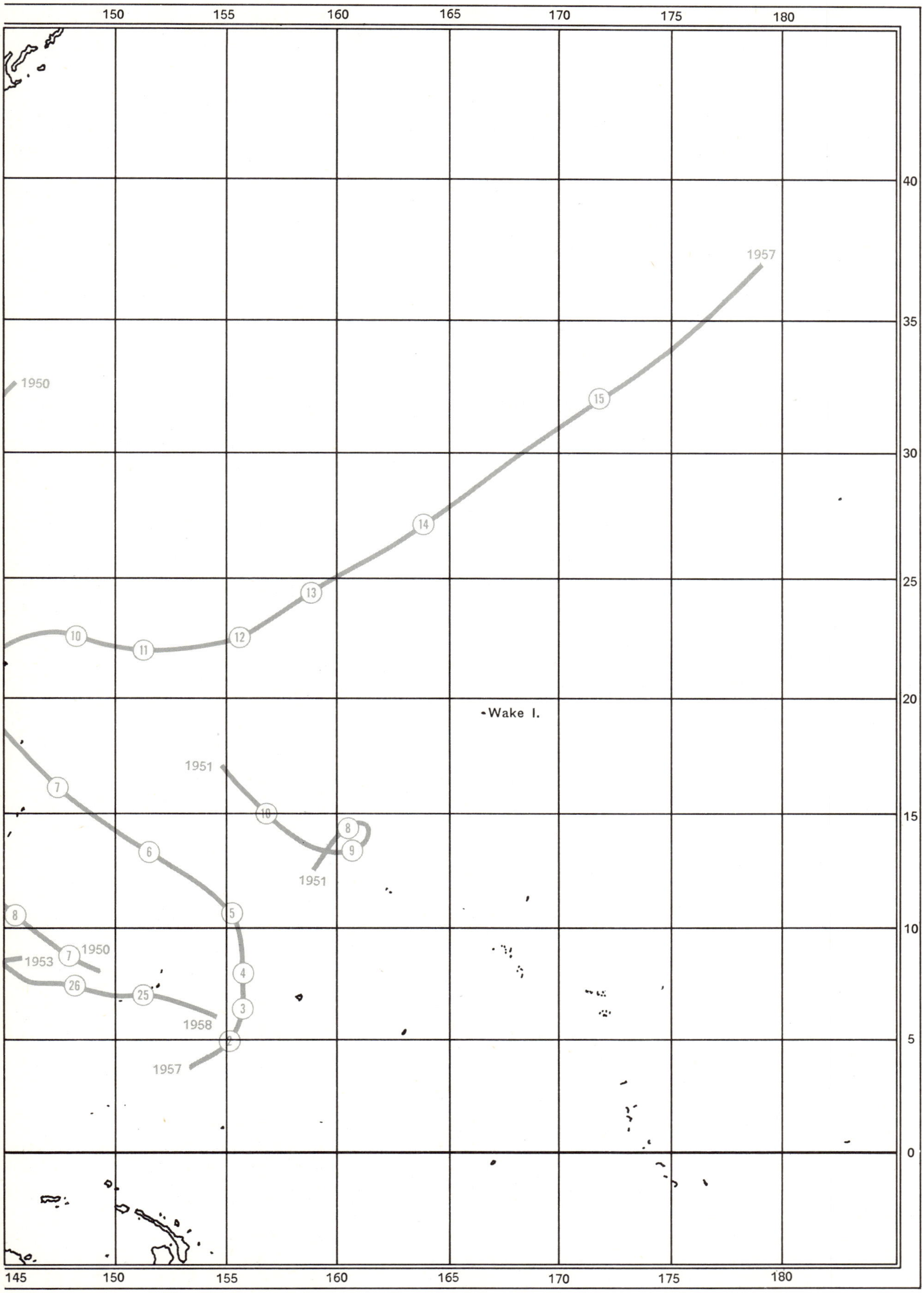

150
155
160
165
170
175
180
40
35
30
25
20
15
10
5
0
1957
15
14
13
12
11
10
1950
Wake I.
1951
10
8
9
1951
7
6
5
4
3
2
1957
8
7
1950
1953
26
25
1958
145

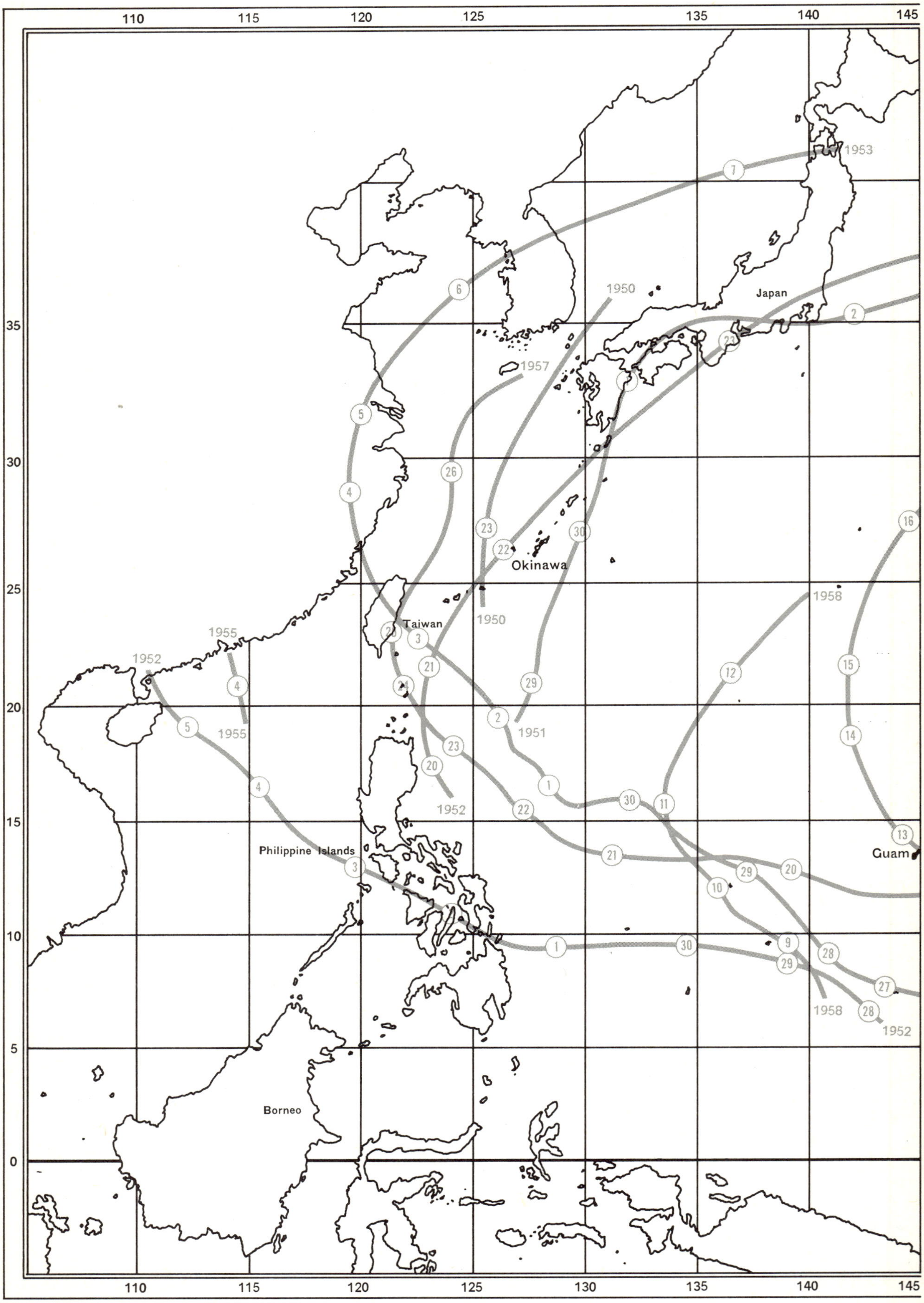

Fig. 18 Typhoon Tracks—June 1950-59. (Circled numbers indicate dates of the month.)

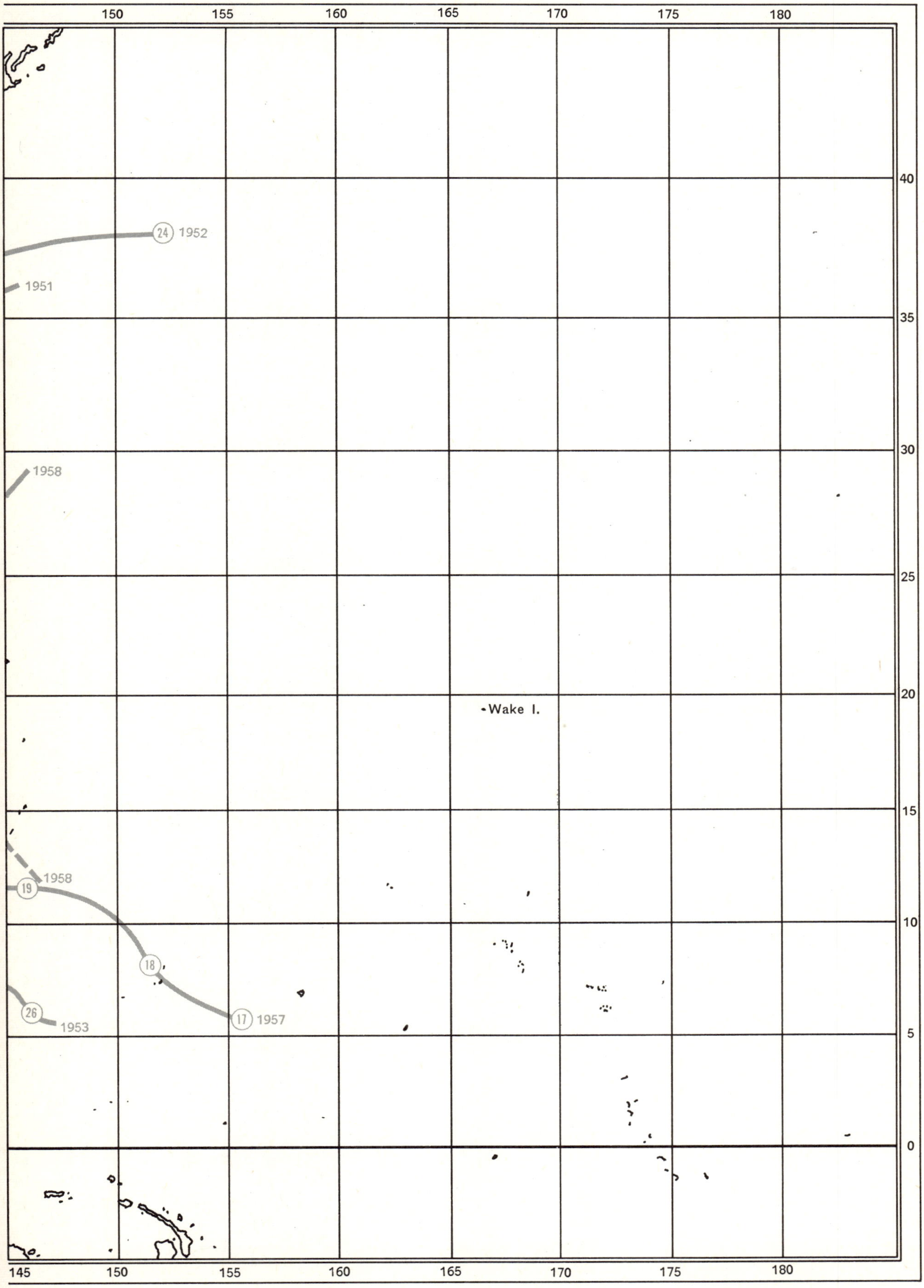
150
155
160
165
170
175
180
40
35
30
25
20
15
10
5
0
24 1952
1951
1958
Wake I.
19 1958
18
26
1953
17 1957
145
150
155
160
165
170
175
180

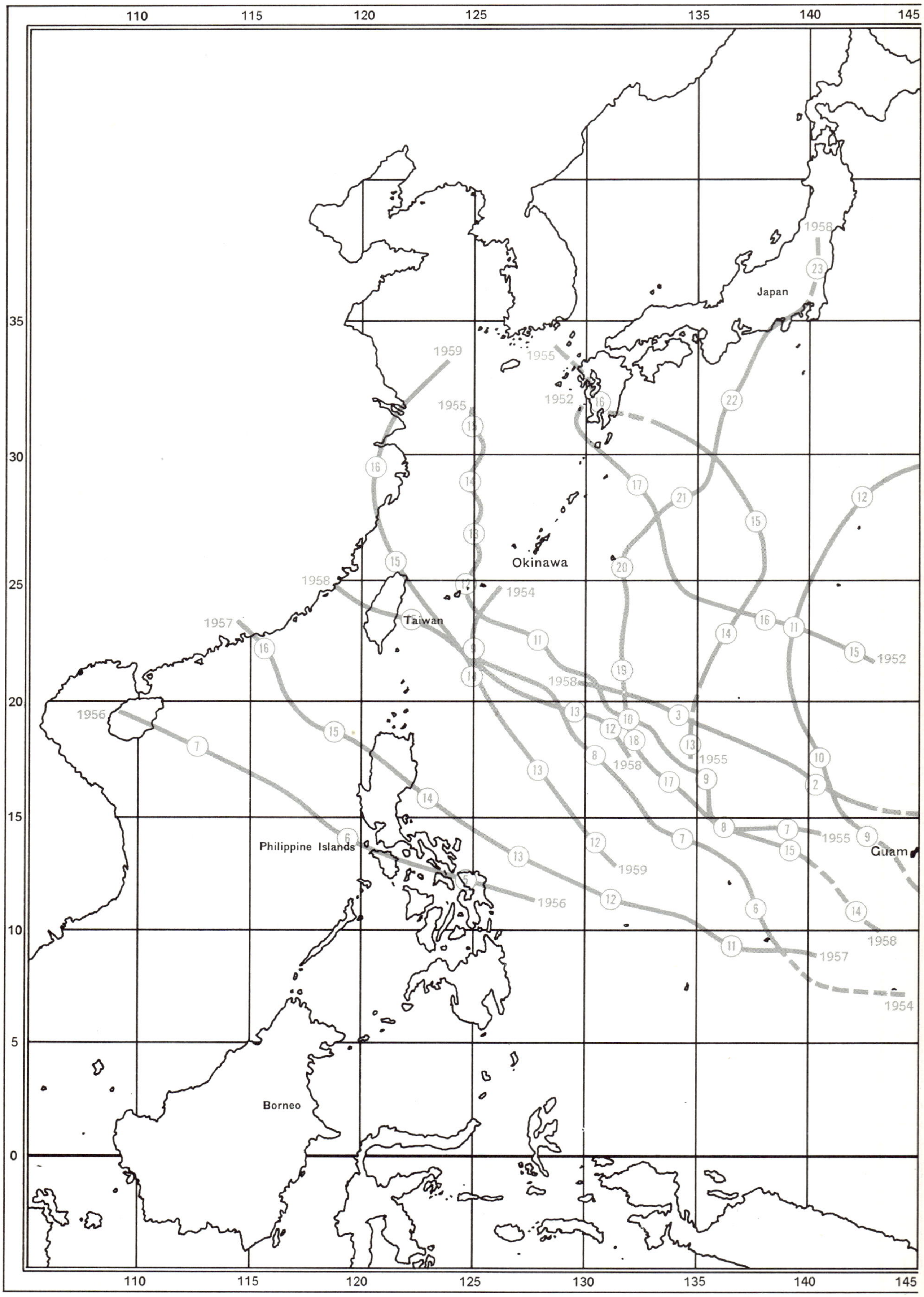

Fig. 19 Typhoon Tracks—1-15 July 1950-59. (Circled numbers indicate dates of the month.)

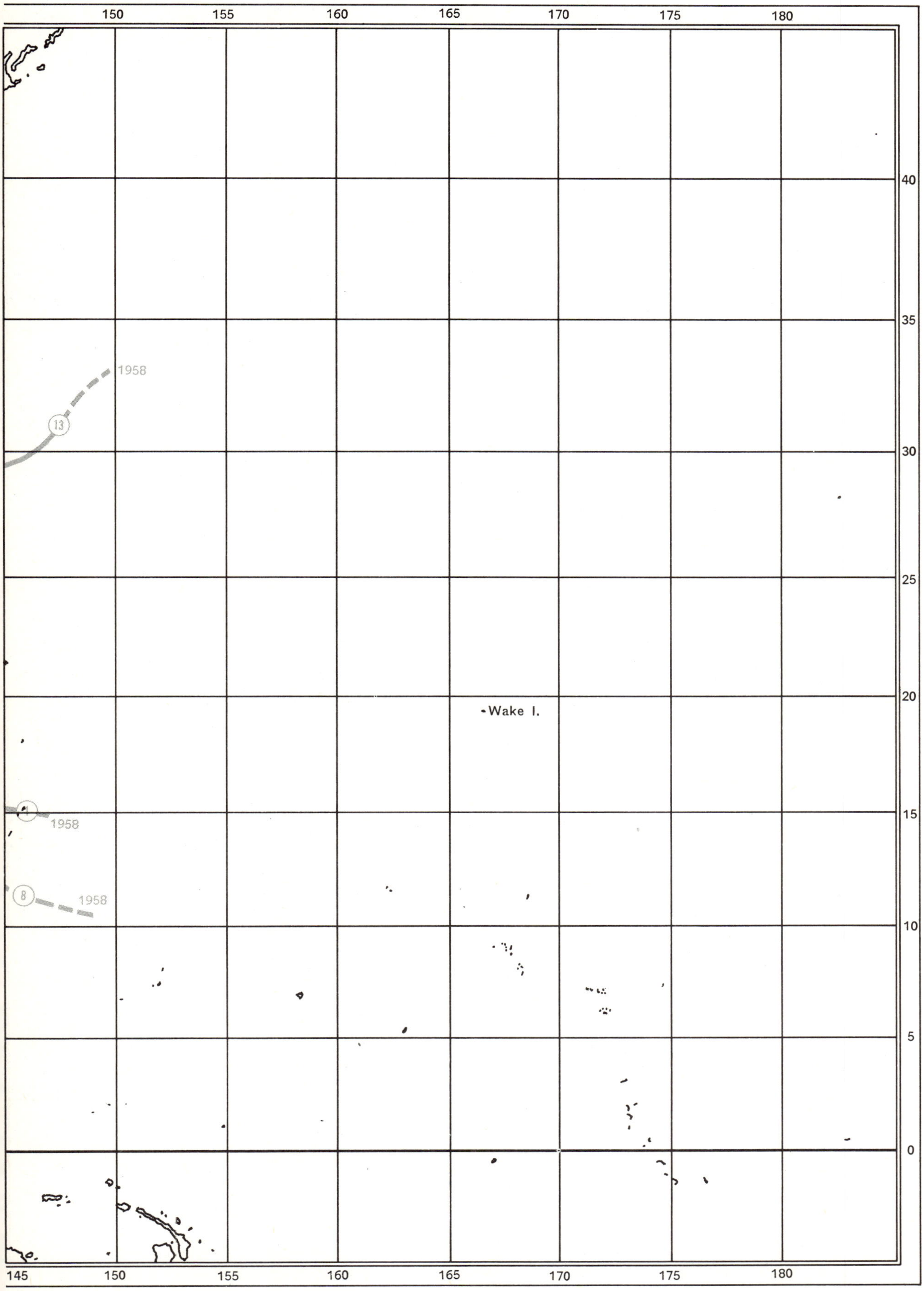
150
155
160
165
170
175
180
40
35
1958
13
30
25
20
Wake I.
15
1958
8
1958
10
5
0
145
150
155
160
165
170
175
180

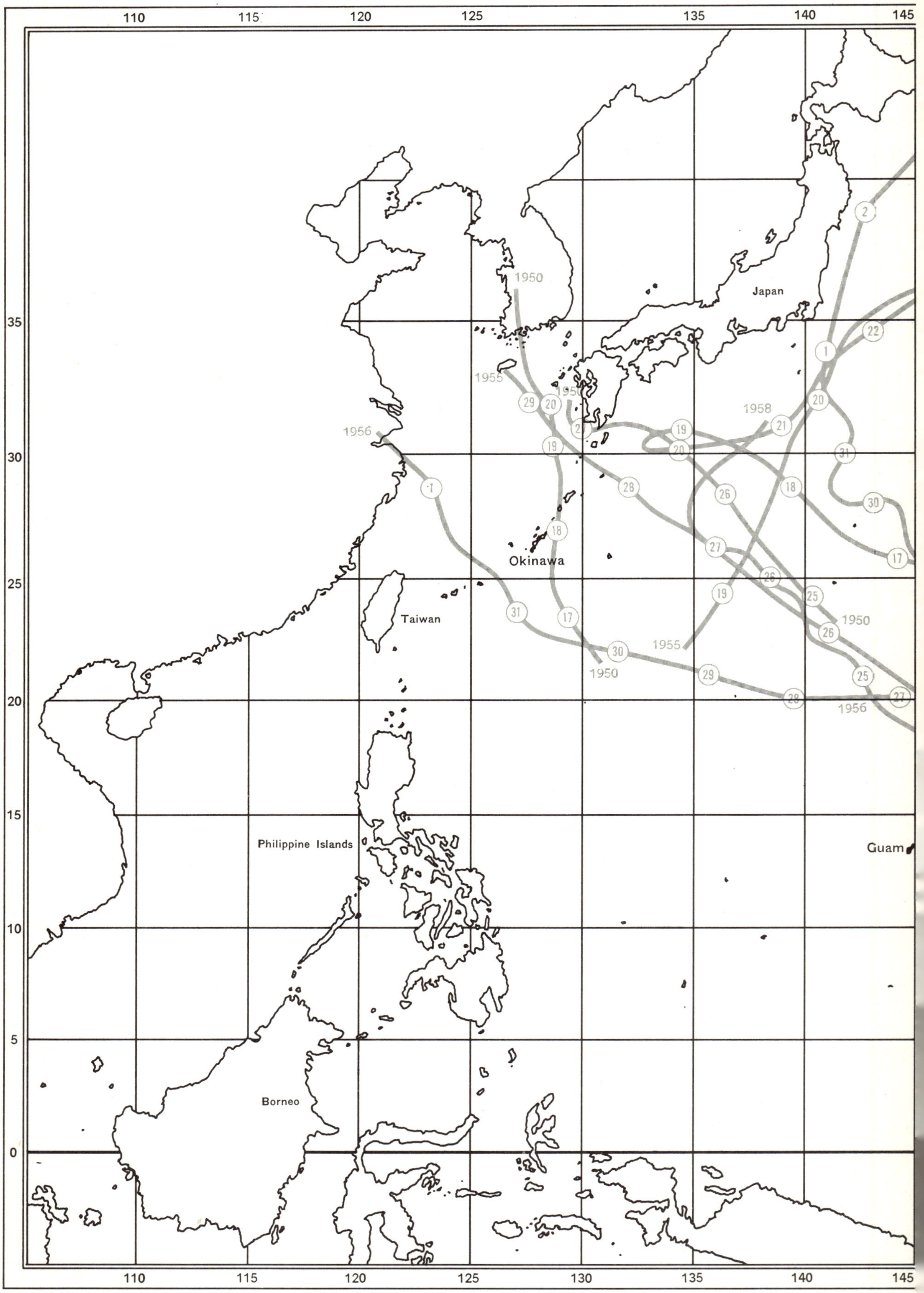

Fig. 20 Typhoon Tracks—16-31 July 1950-59. (Circled numbers indicate dates of the month.)

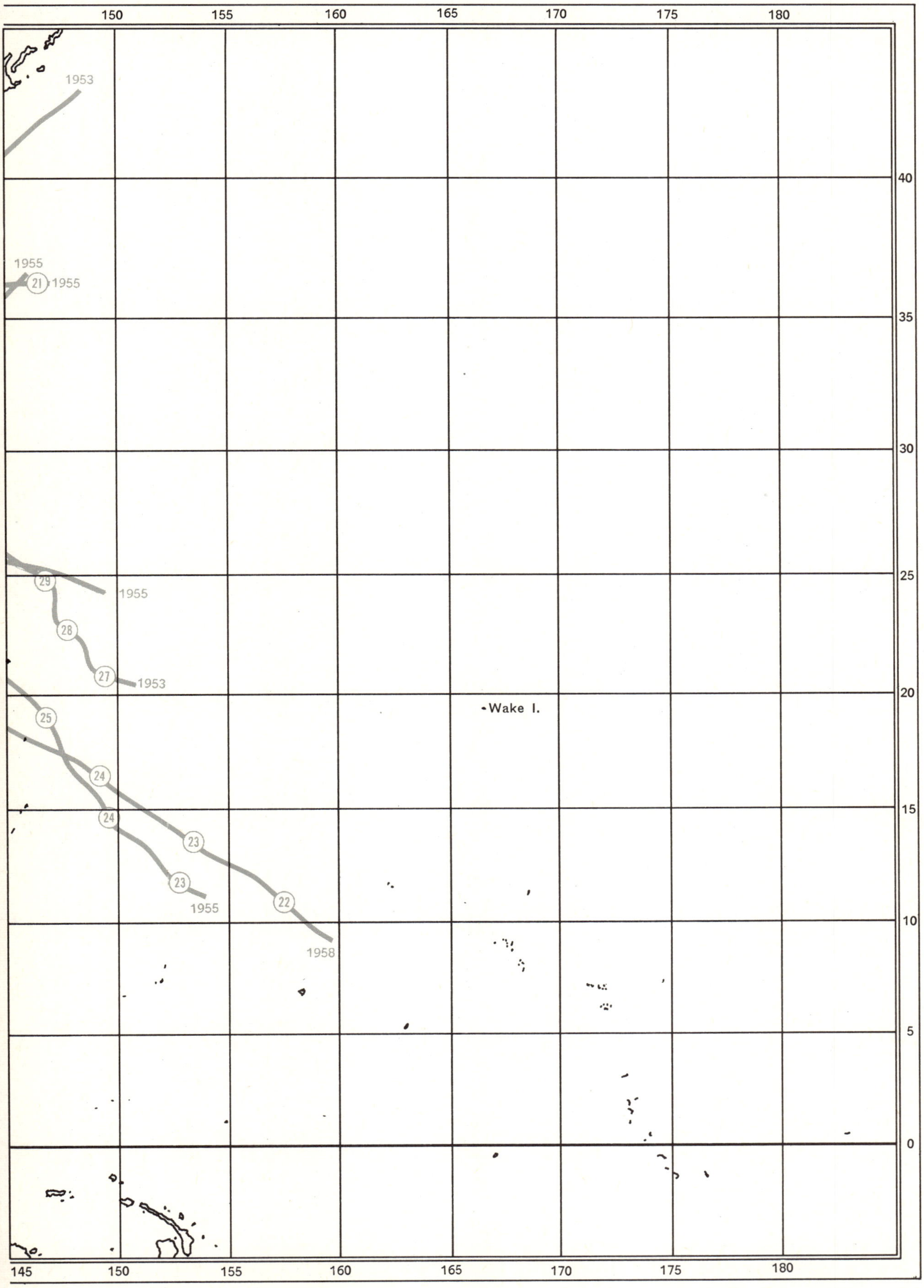
150
155
160
165
170
175
180
1953
40
1955
21
1955
35
30
25
29
1955
28
27
1953
20
Wake I.
25
24
15
24
23
23
1955
22
10
1958
5
0
145
150
155
160
165
170
175
180

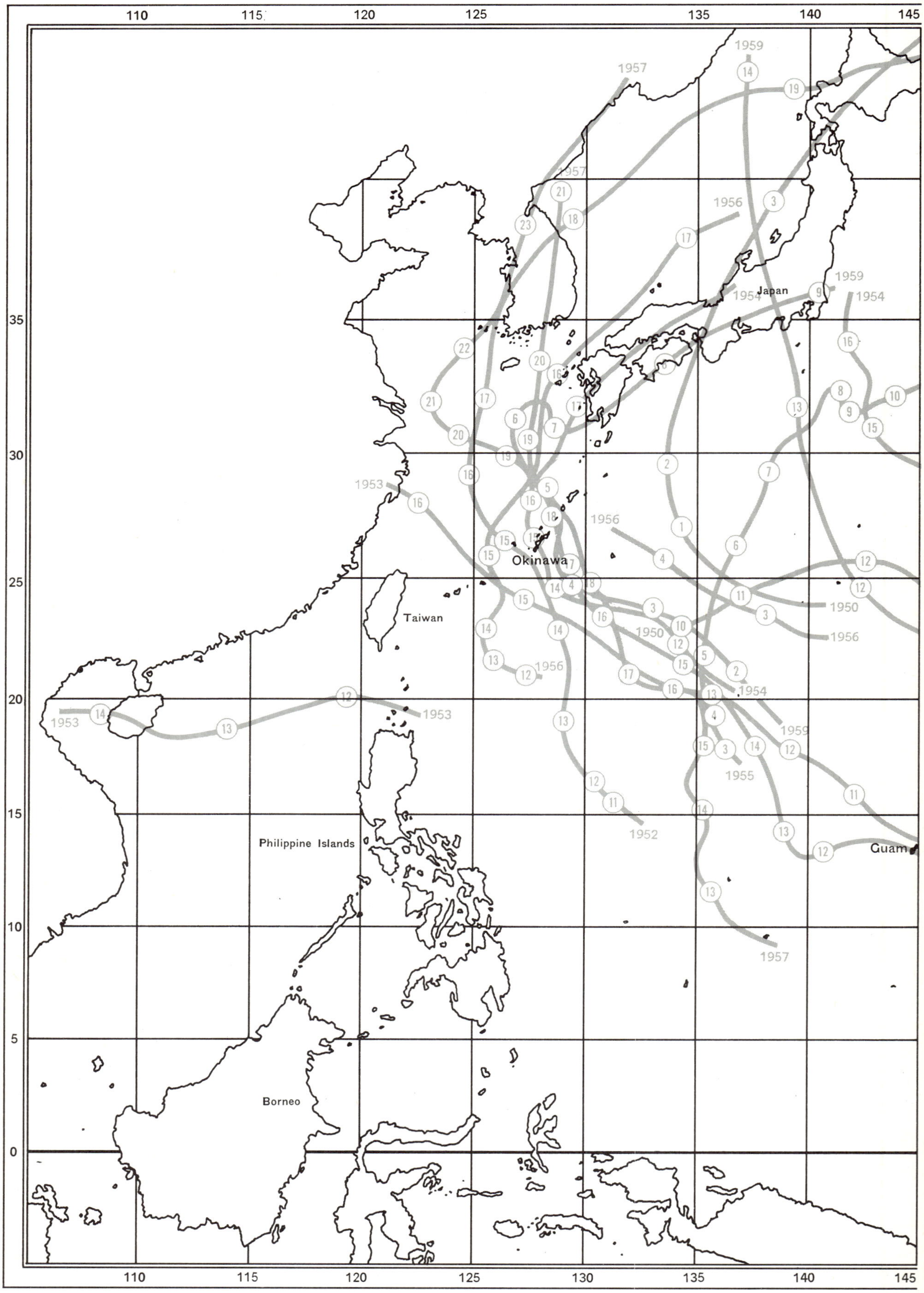

Fig. 21 Typhoon Tracks—1-15 August 1950-59. (Circled numbers indicate dates of the month.)

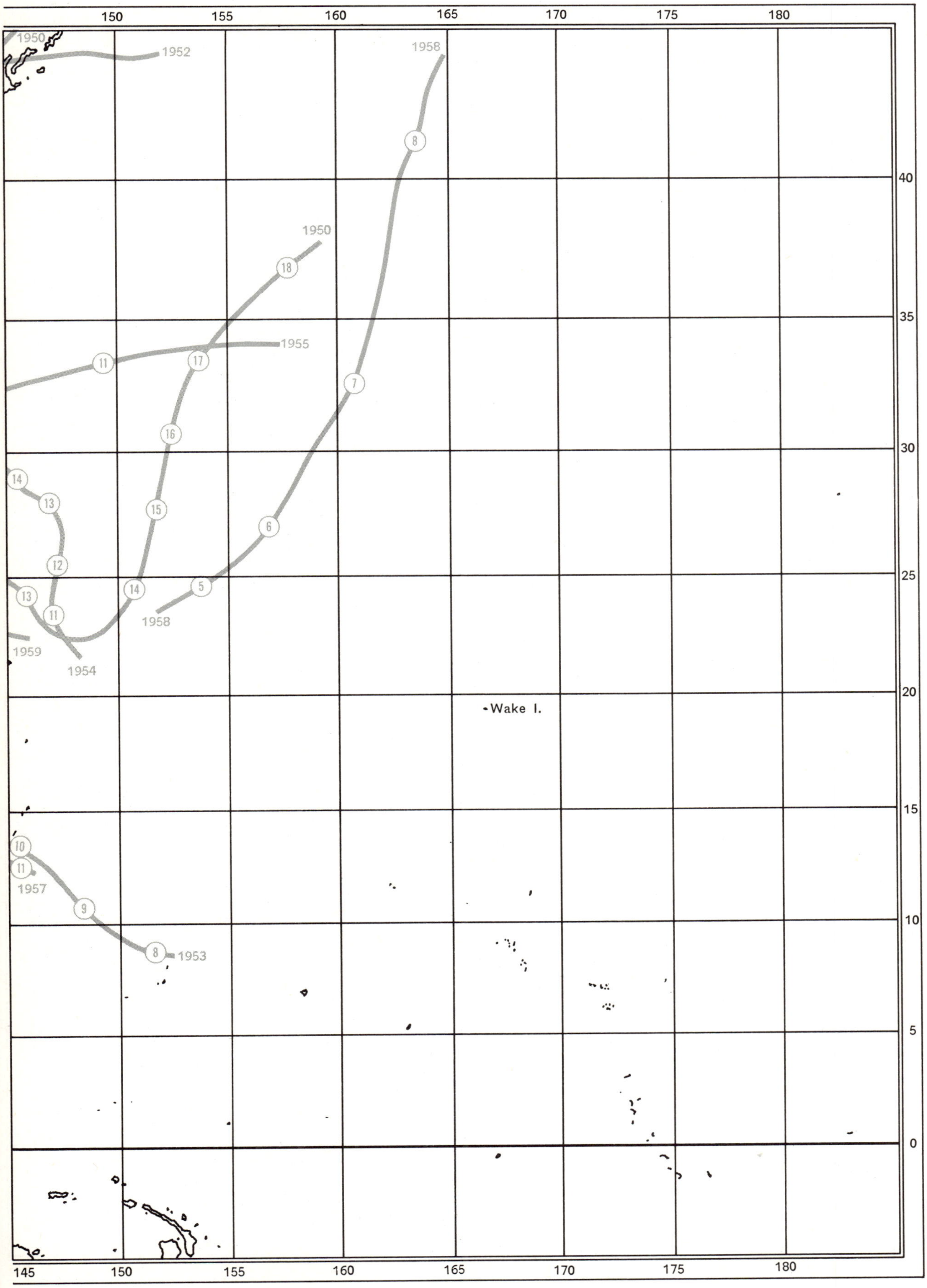
150
155
160
165
170
175
180
1950
1952
1958
8
40
1950
18
35
1955
11
17
7
16
30
14
13
15
6
12
25
13
14
5
11
1958
1959
1954
20
Wake I.
15
10
11
1957
9
10
8
1953
5
0
145
150
155
160
165
170
175
180

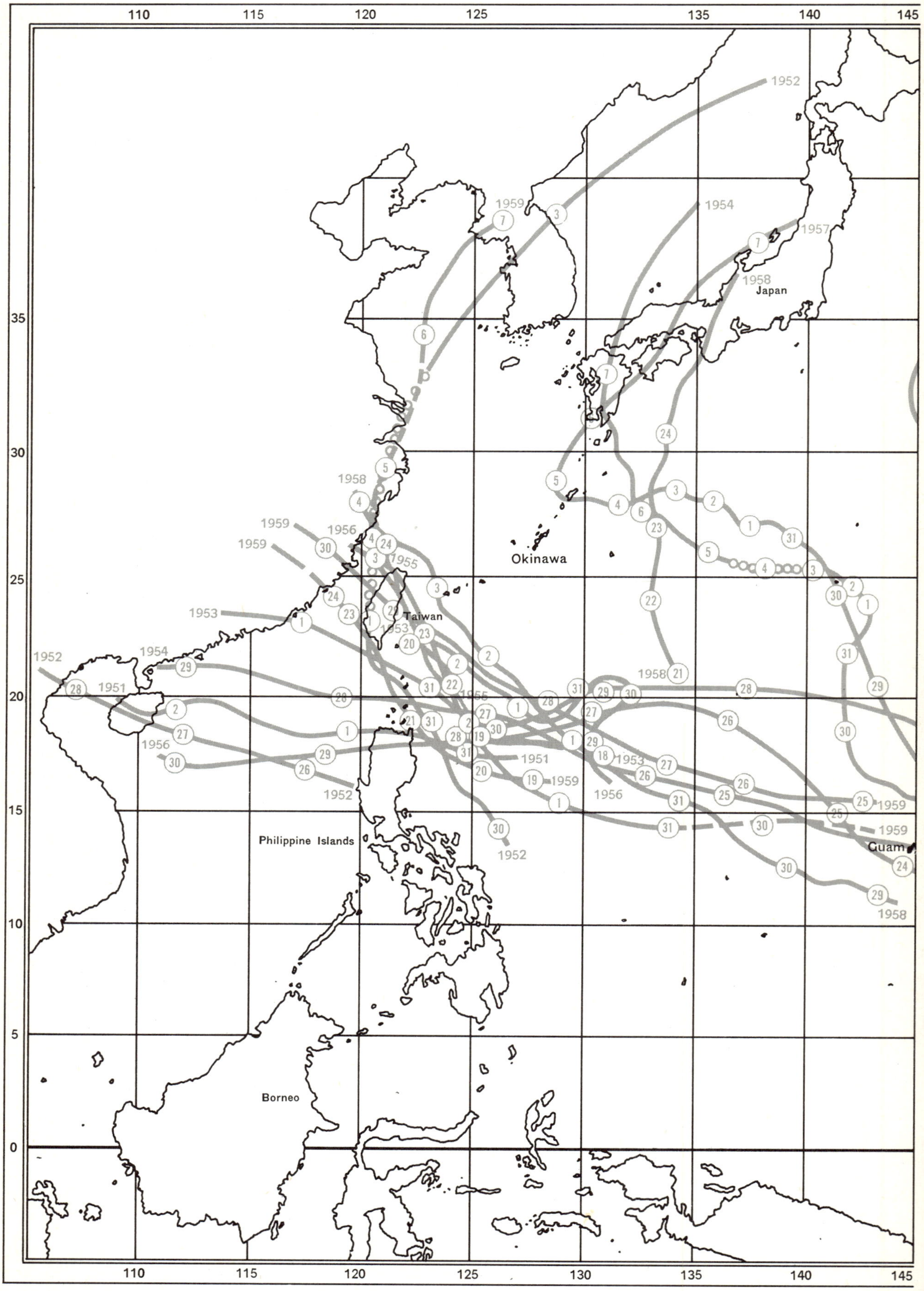

Fig. 22 Typhoon Tracks—16-31 August 1950-59. (Circled numbers indicate dates of the month.)

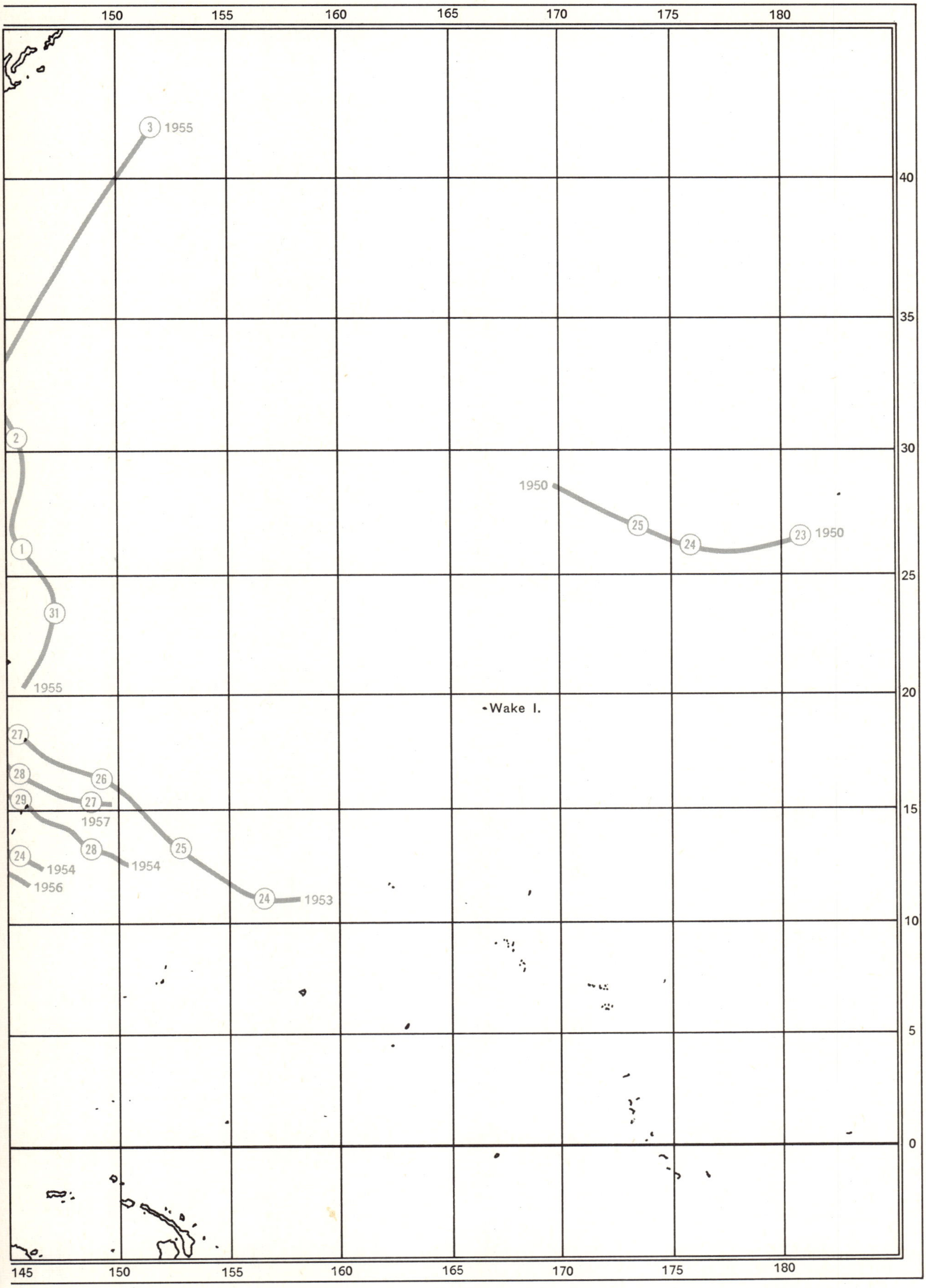
150
155
160
165
170
175
180
40
35
30
25
20
15
10
5
0
145
3 1955
2
1
31
1955
1950
25
24
23 1950
Wake I.
27
28
29
26
27
1957
28
25
24 1954
1954
1956
24
1953

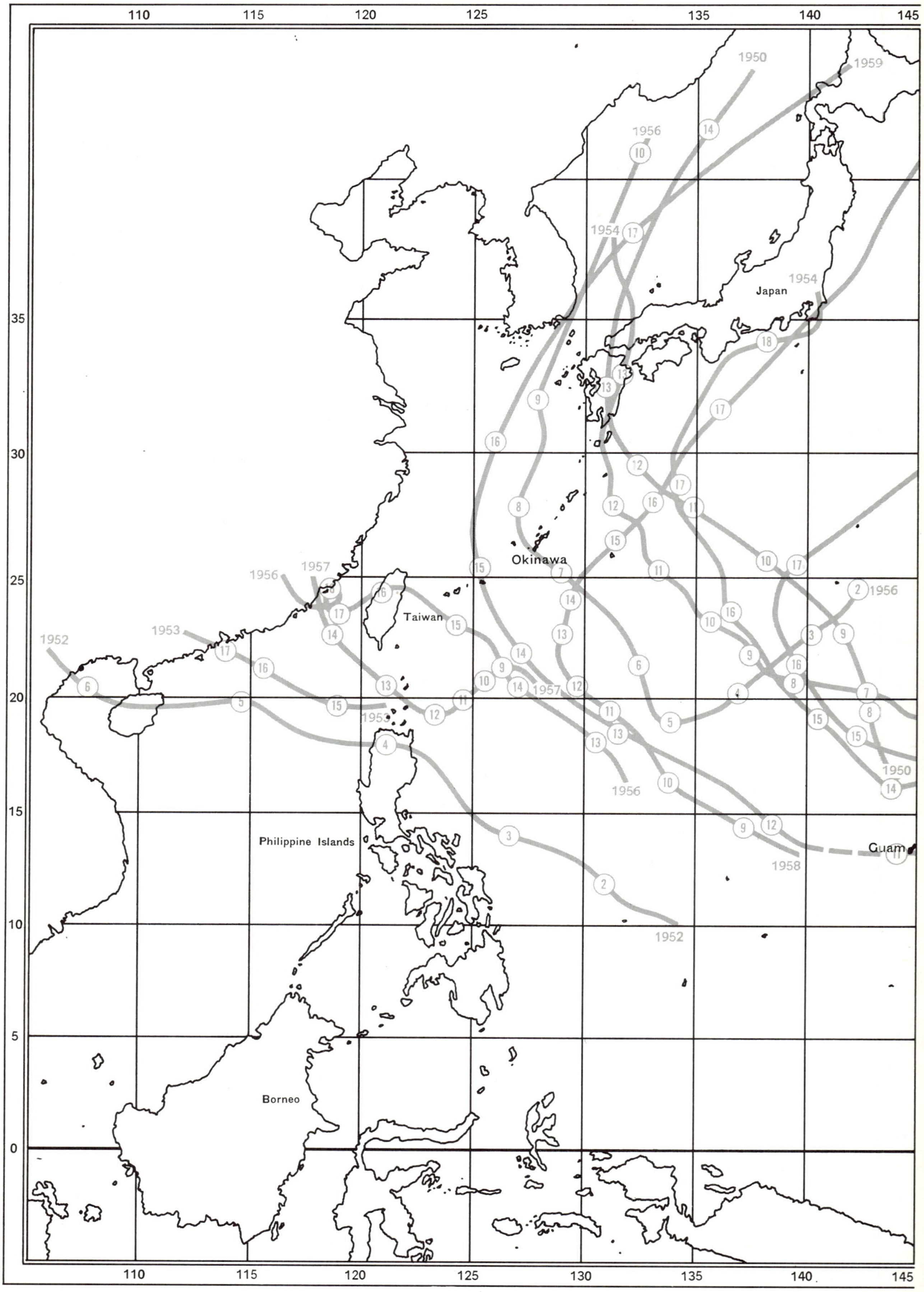

Fig. 23 Typhoon Tracks—1-15 September 1950-59. (Circled numbers indicate dates of the month.)

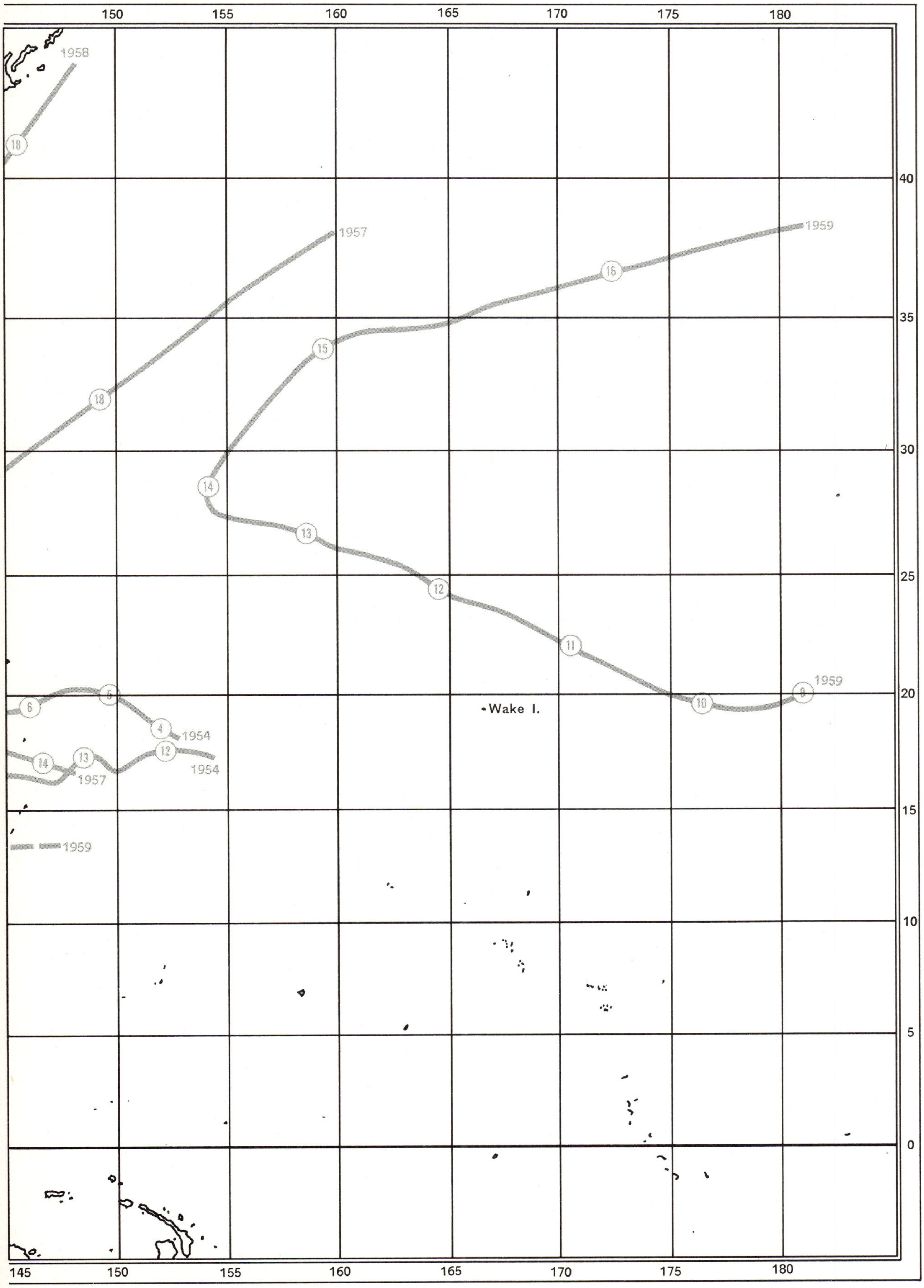
150
155
160
165
170
175
180
1958
18
40
1957
1959
16
35
15
18
30
14
13
25
12
11
20
1959
6
5
10
Wake I.
4
1954
14
13
12
1957
1954
15
1959
10
5
0
145
150
155
160
165
170
175
180

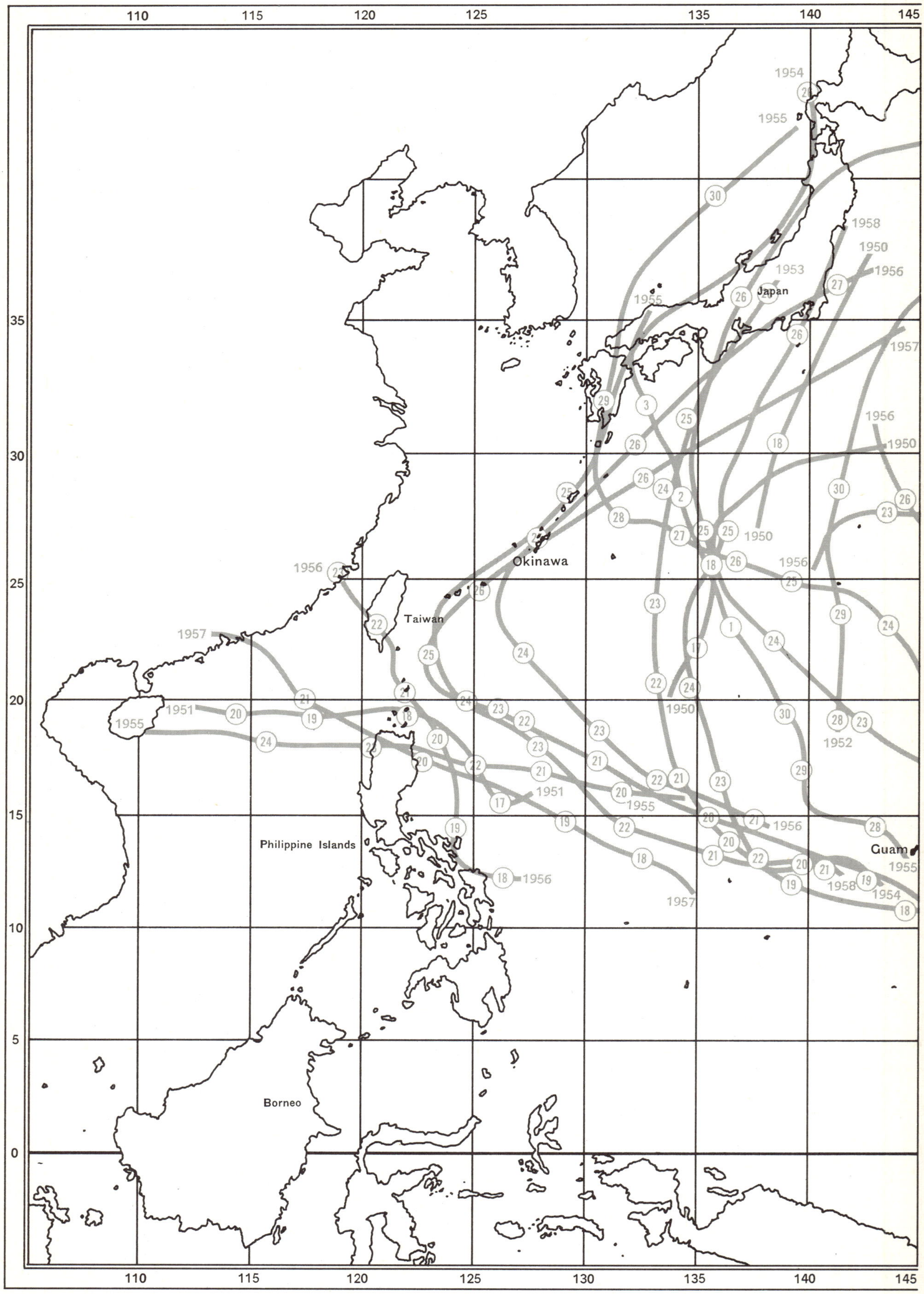

Fig. 24 Typhoon Tracks—16-30 September 1950-59. (Circled numbers indicate dates of the month.)

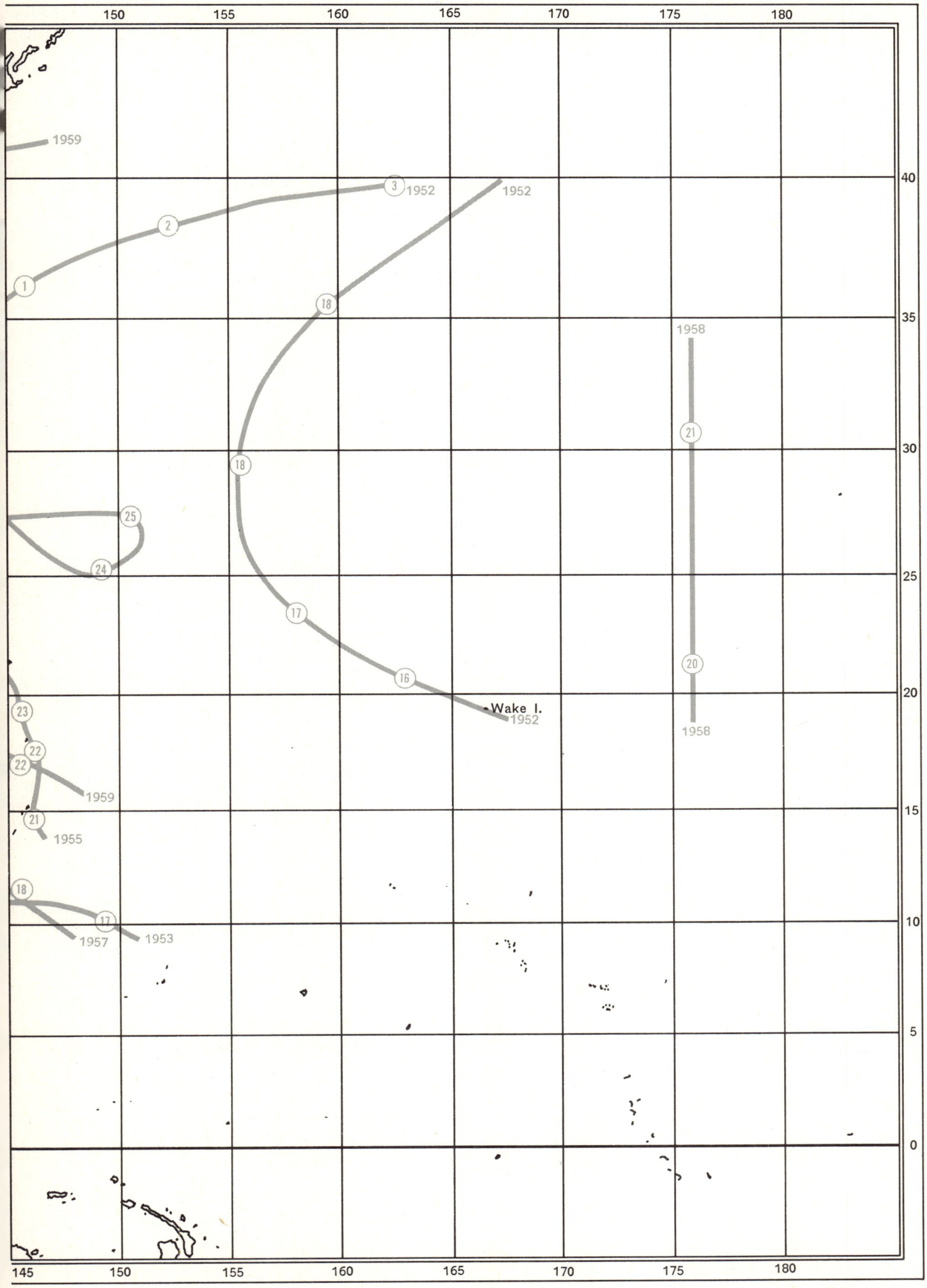

150
155
160
165
170
175
180
1959
1952
1952
1958
1958
1952
Wake I.
1959
1955
1957
1953
40
35
30
25
20
15
10
5
0
145

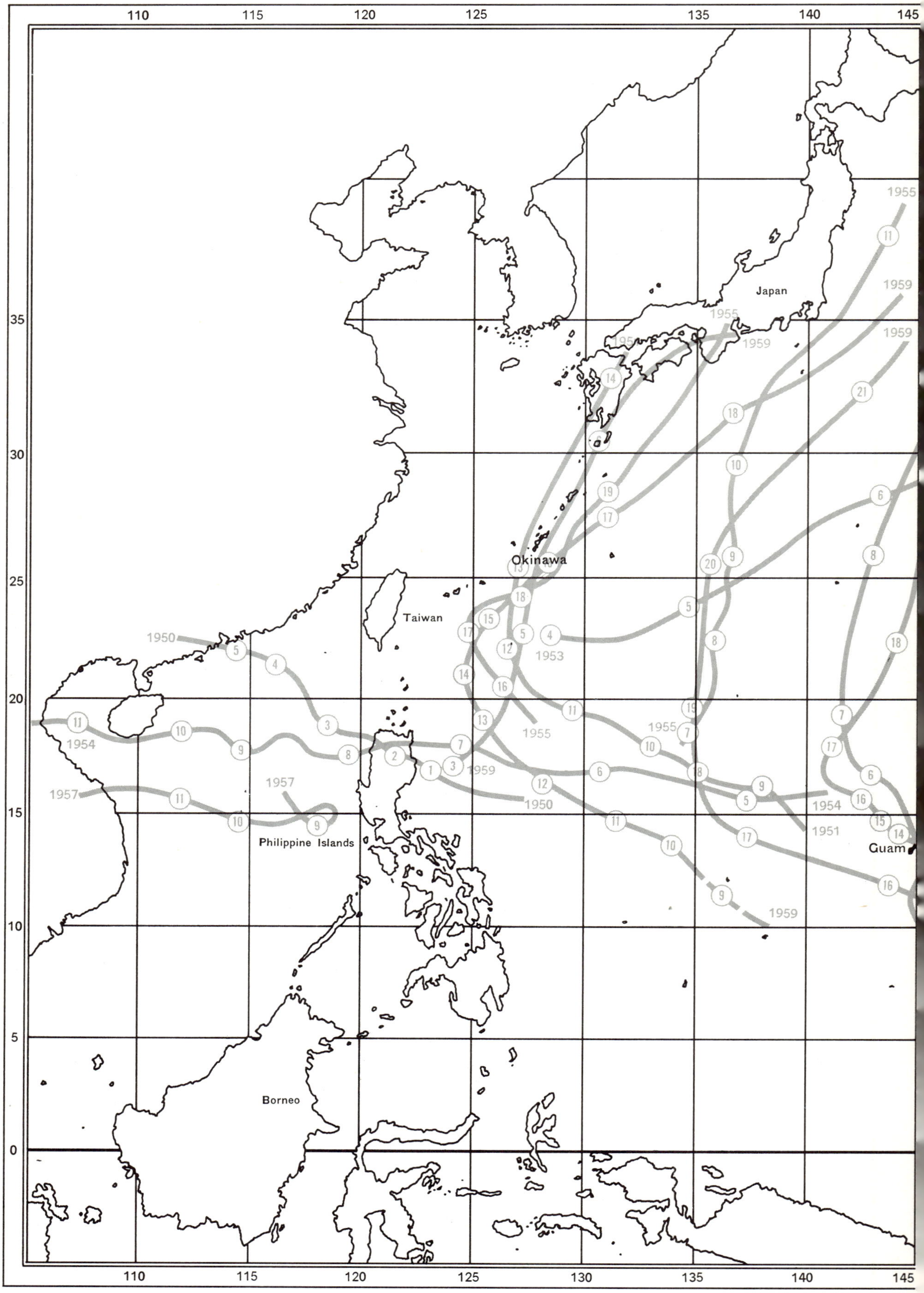

Fig. 25 Typhoon Tracks—1-15 October 1950-59. (Circled numbers indicate dates of the month.)

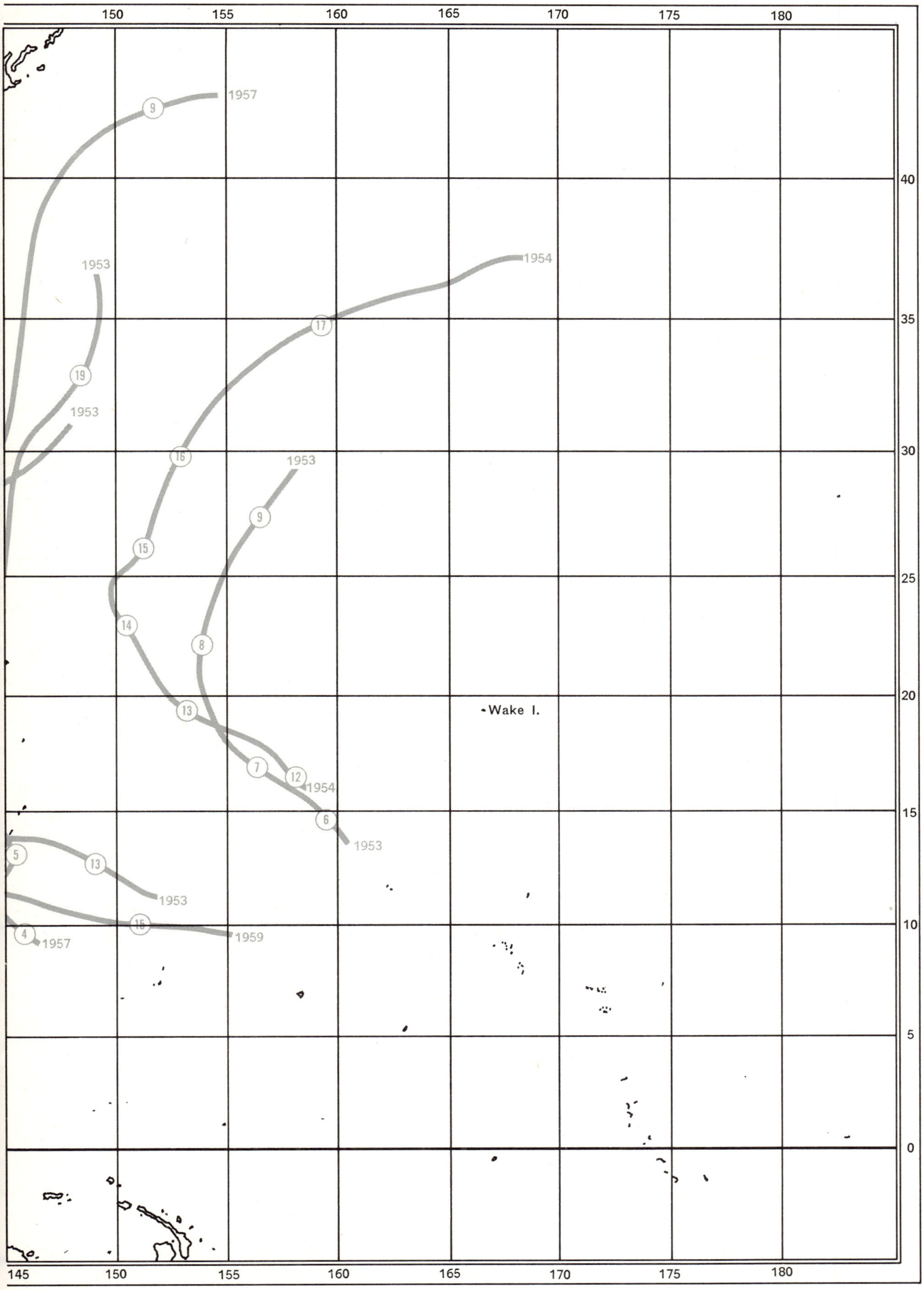

150
155
160
165
170
175
180
1957
9
40
1954
1953
35
17
19
1953
30
16
1953
9
15
25
14
8
20
Wake I.
13
7
12
1954
15
6
1953
5
13
1953
10
16
4
1957
1959
5
0
145
150
155
160
165
170
175
180

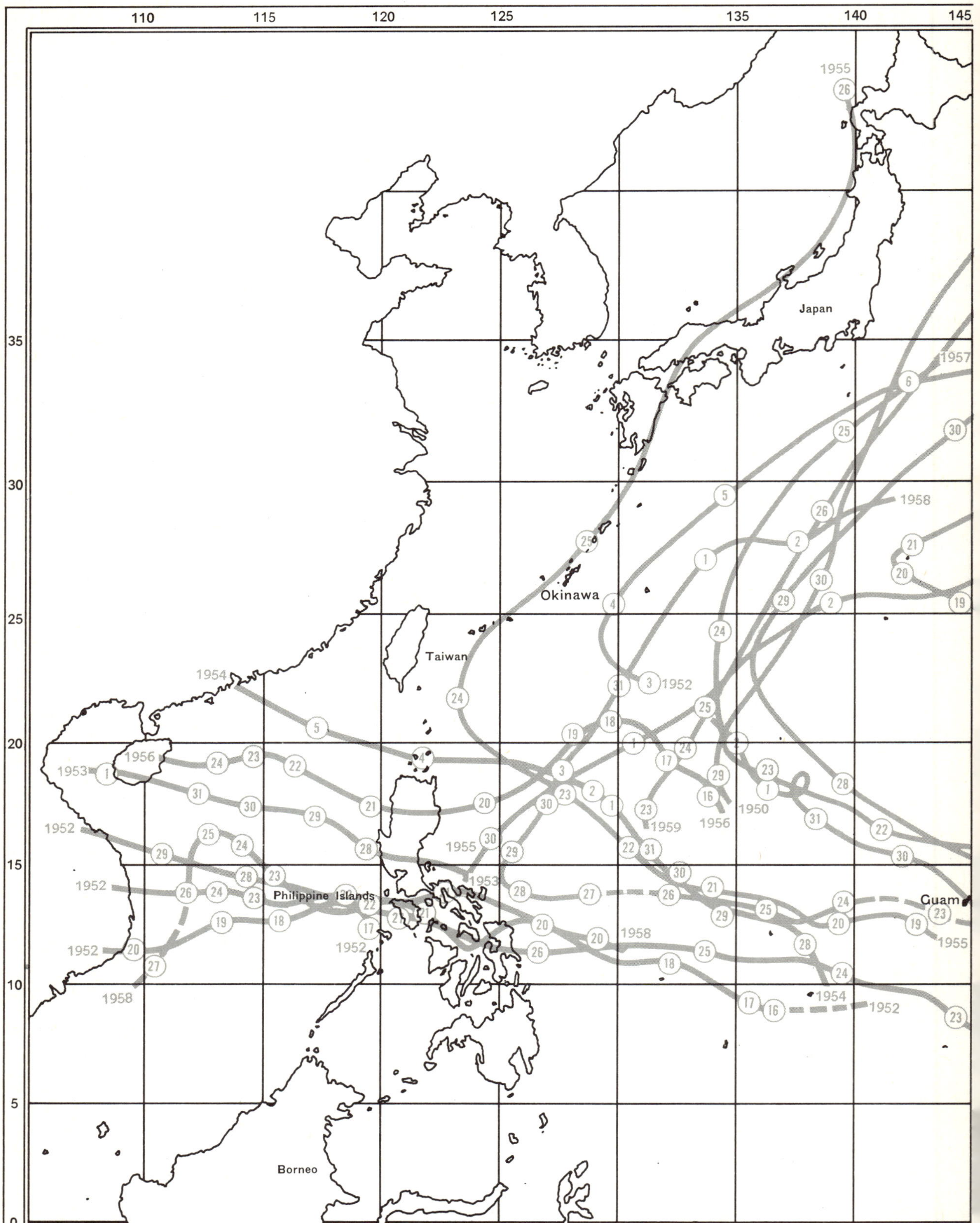

Fig. 26 Typhoon Tracks—16-31 October 1950-59. (Circled numbers indicate dates of the month.)

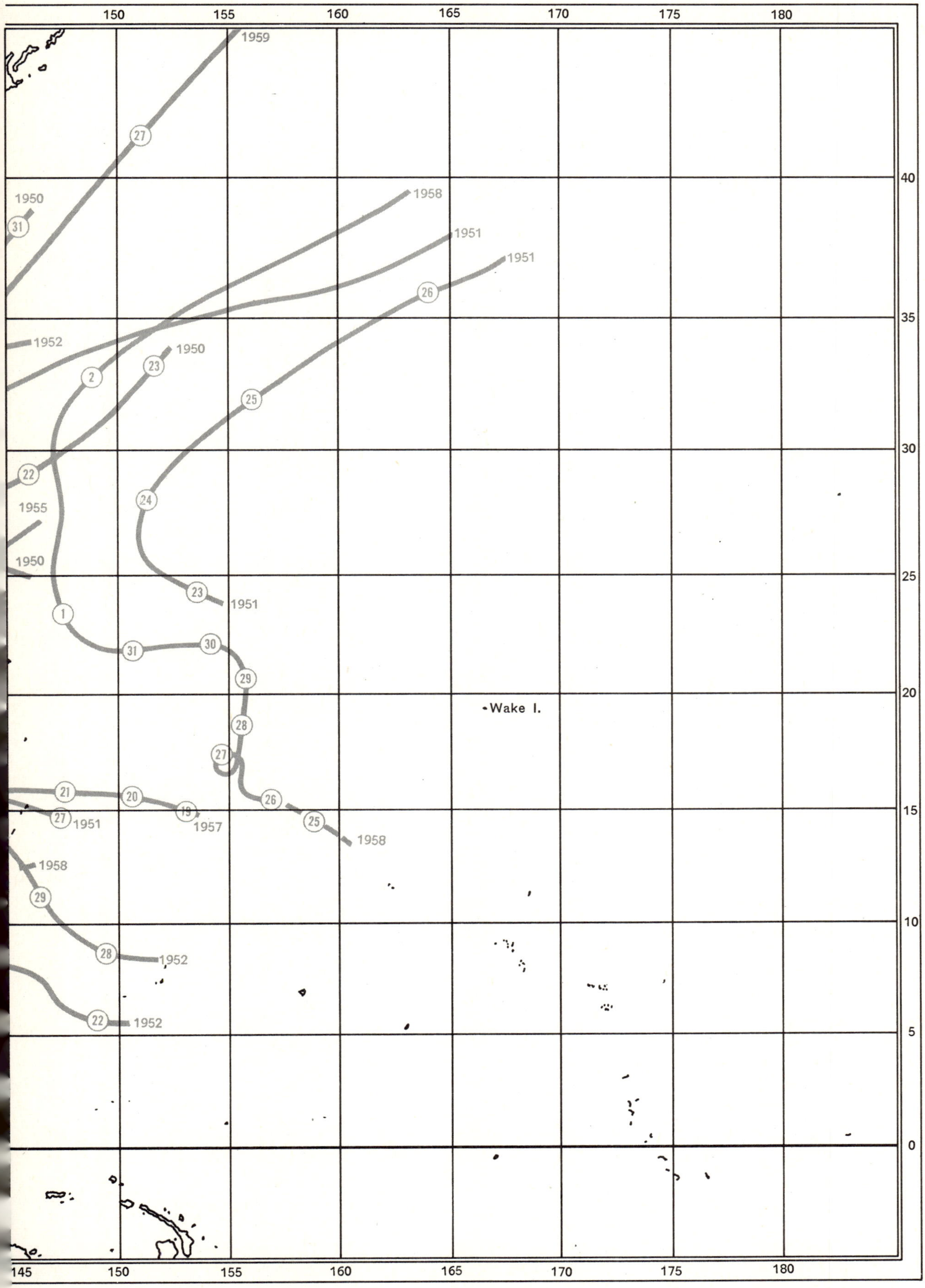

150
155
160
165
170
175
180
1959
27
1950
31
1958
1951
1951
26
1952
1950
2
23
25
22
24
1955
1950
1
23
1951
31
30
29
Wake I.
28
27
21
20
19
26
25
27
1951
1957
1958
1958
29
28
1952
22
1952
40
35
30
25
20
15
10
5
0
145
150
155
160
165
170
175
180

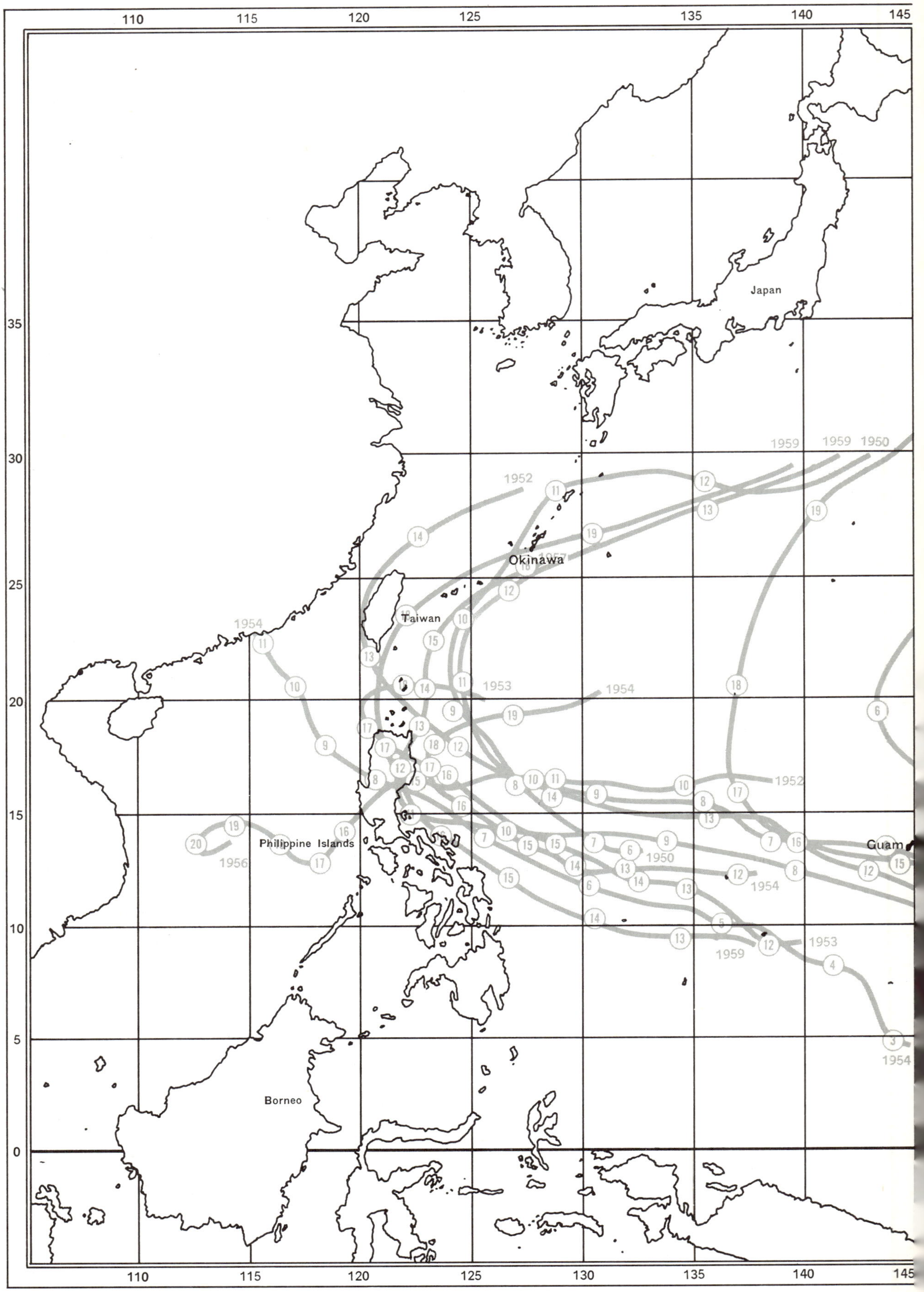

Fig. 27 Typhoon Tracks—1-15 November 1950-59. (Circled numbers indicate dates of the month.)

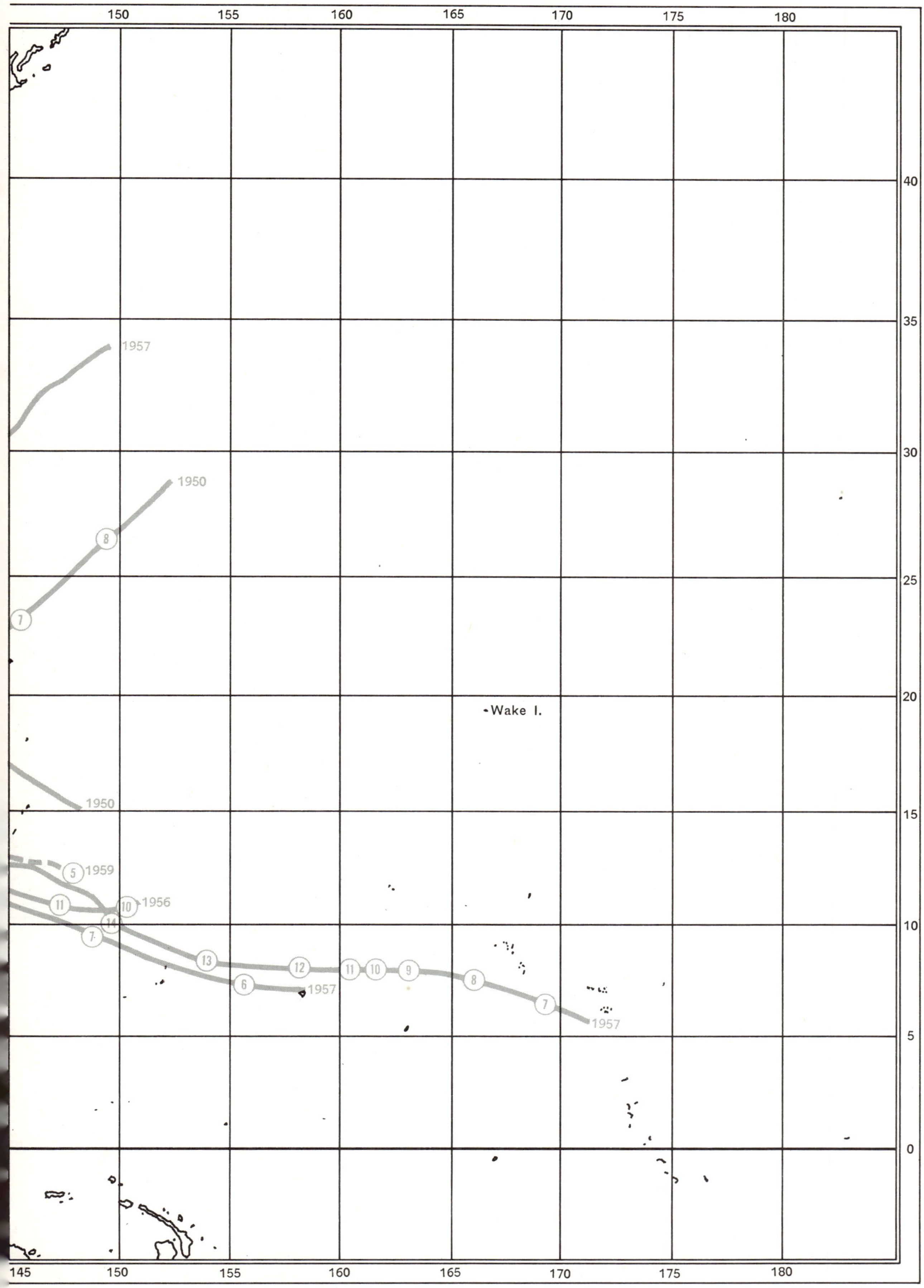
150
155
160
165
170
175
180
40
35
30
25
20
15
10
5
0
1957
1950
Wake I.
1950
1959
1956
1957
1957
145

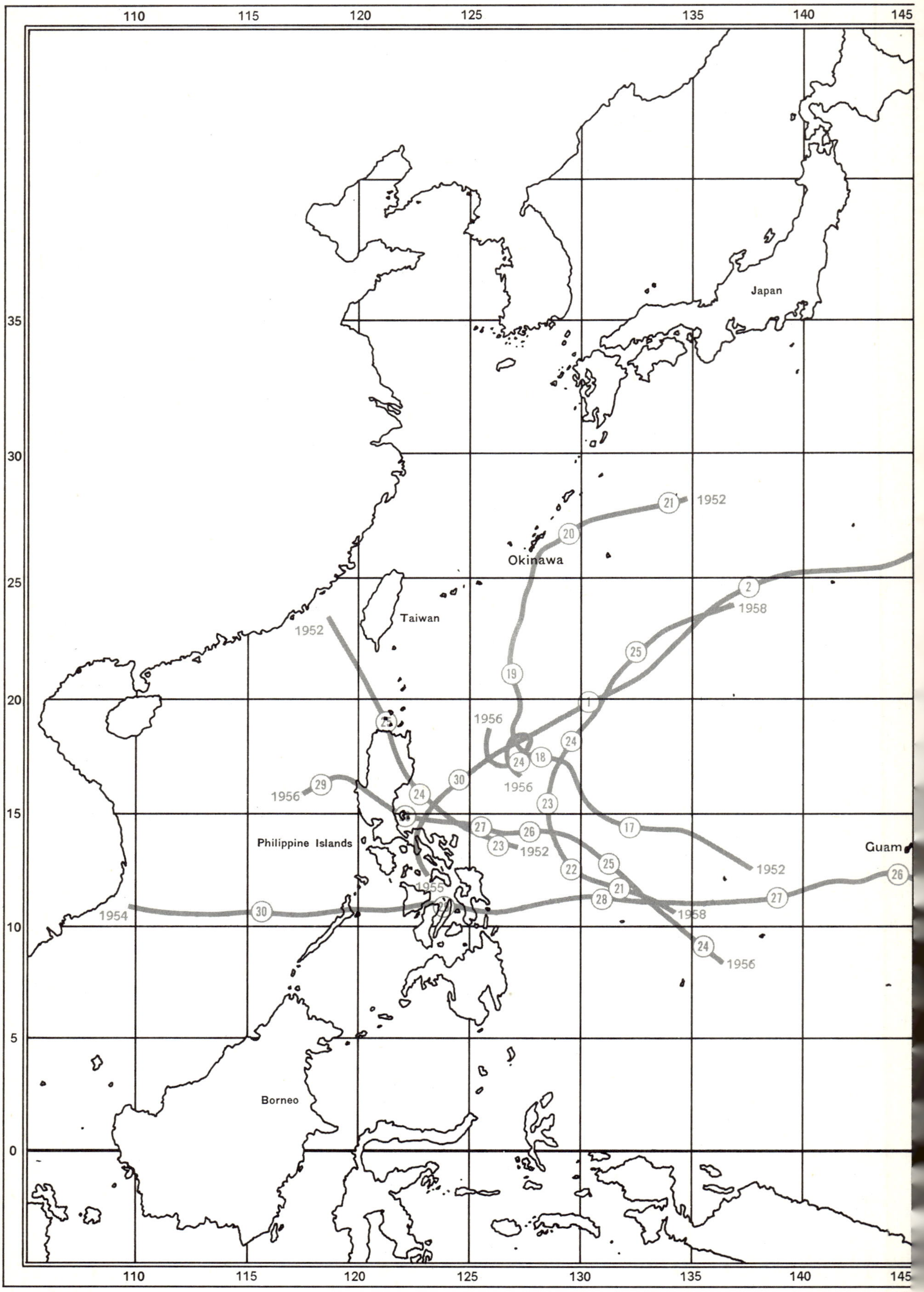

Fig. 28 ***Typhoon Tracks—16-30 November 1950-59.*** *(Circled numbers indicate dates of the month.)*

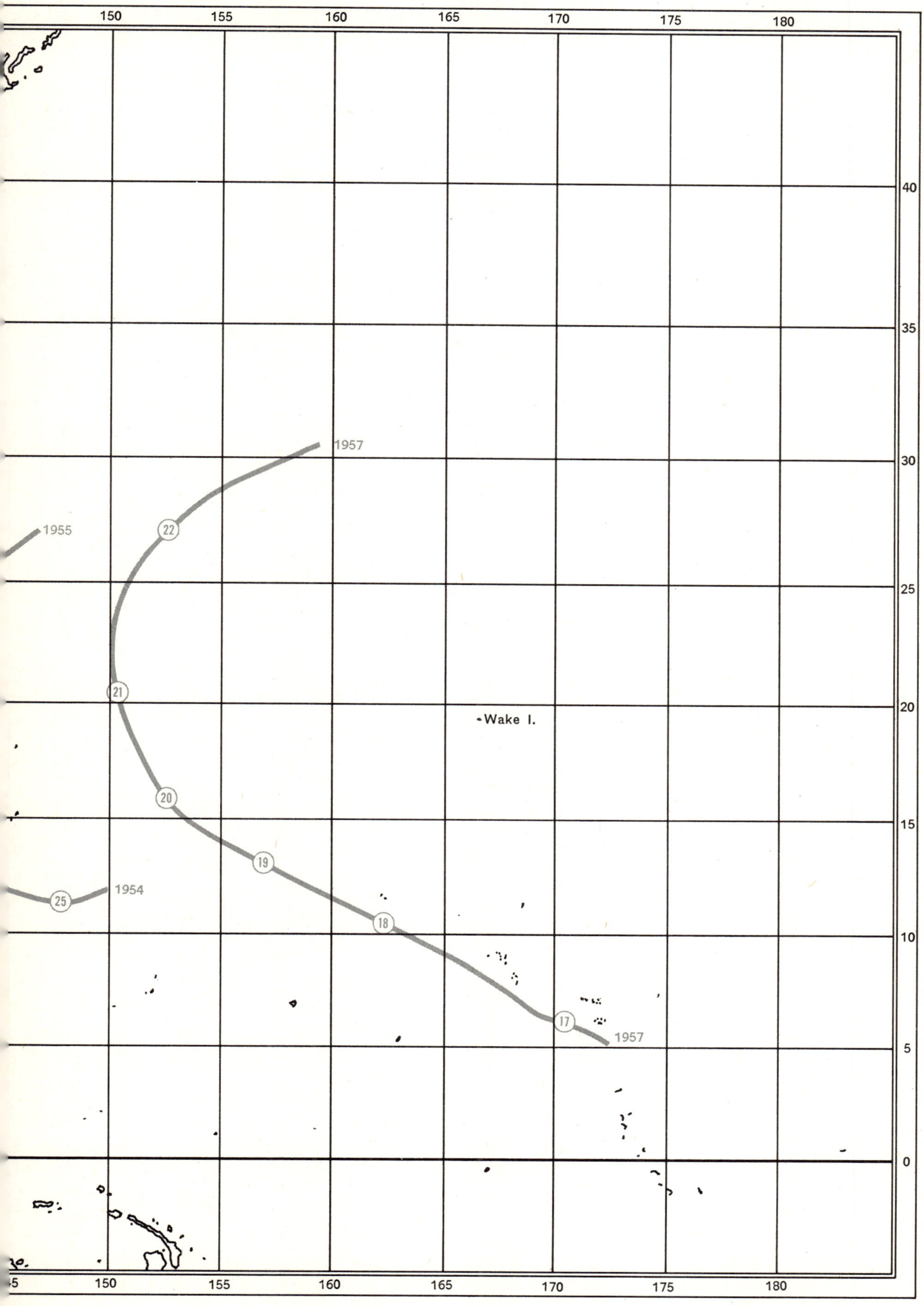
150
155
160
165
170
175
180
40
35
30
25
20
15
10
5
0
1957
1955
1954
22
21
20
19
18
17
25
Wake I.

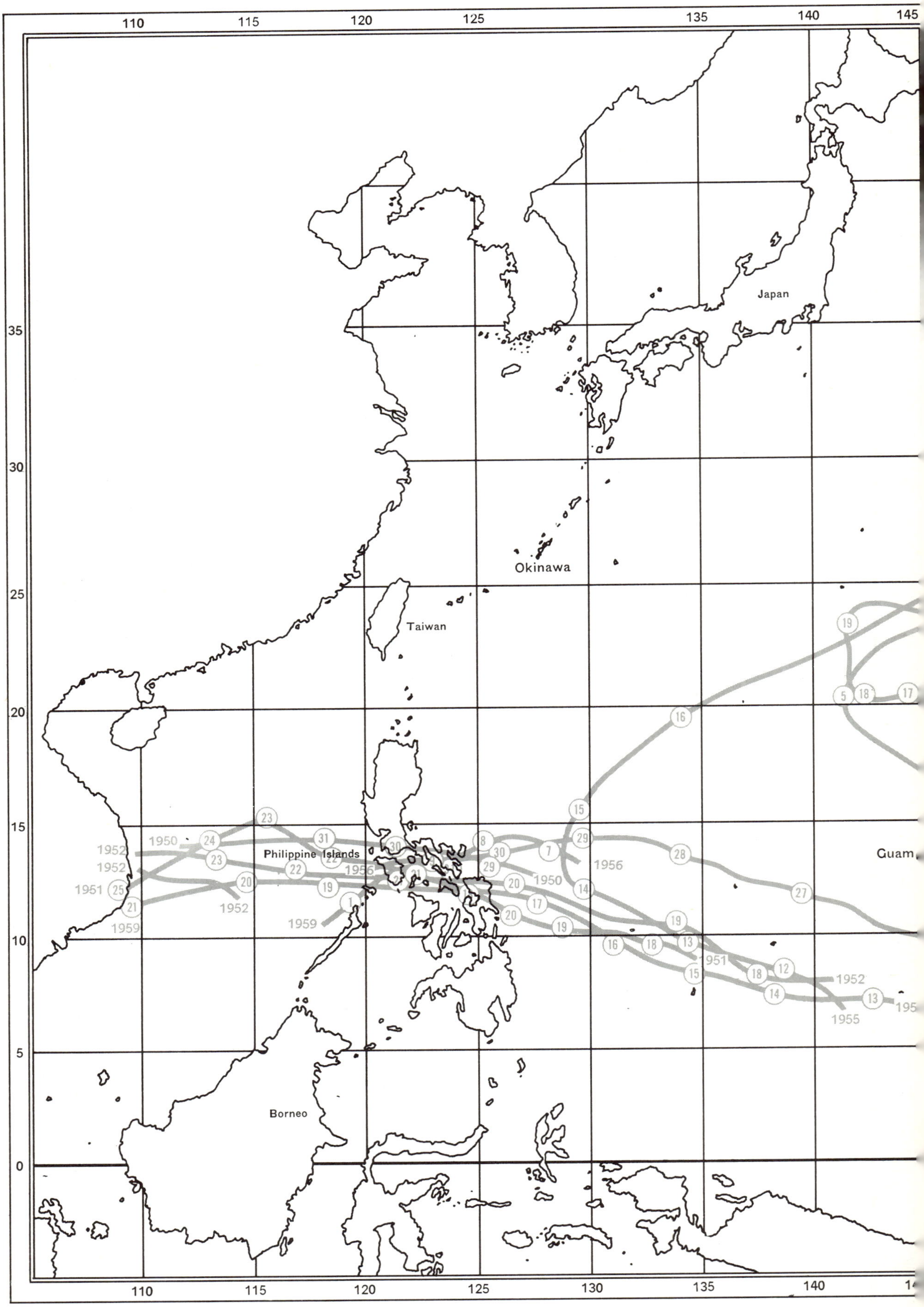

Fig. 29 Typhoon Tracks—December 1950-59. (Circled numbers indicate dates of the month.)

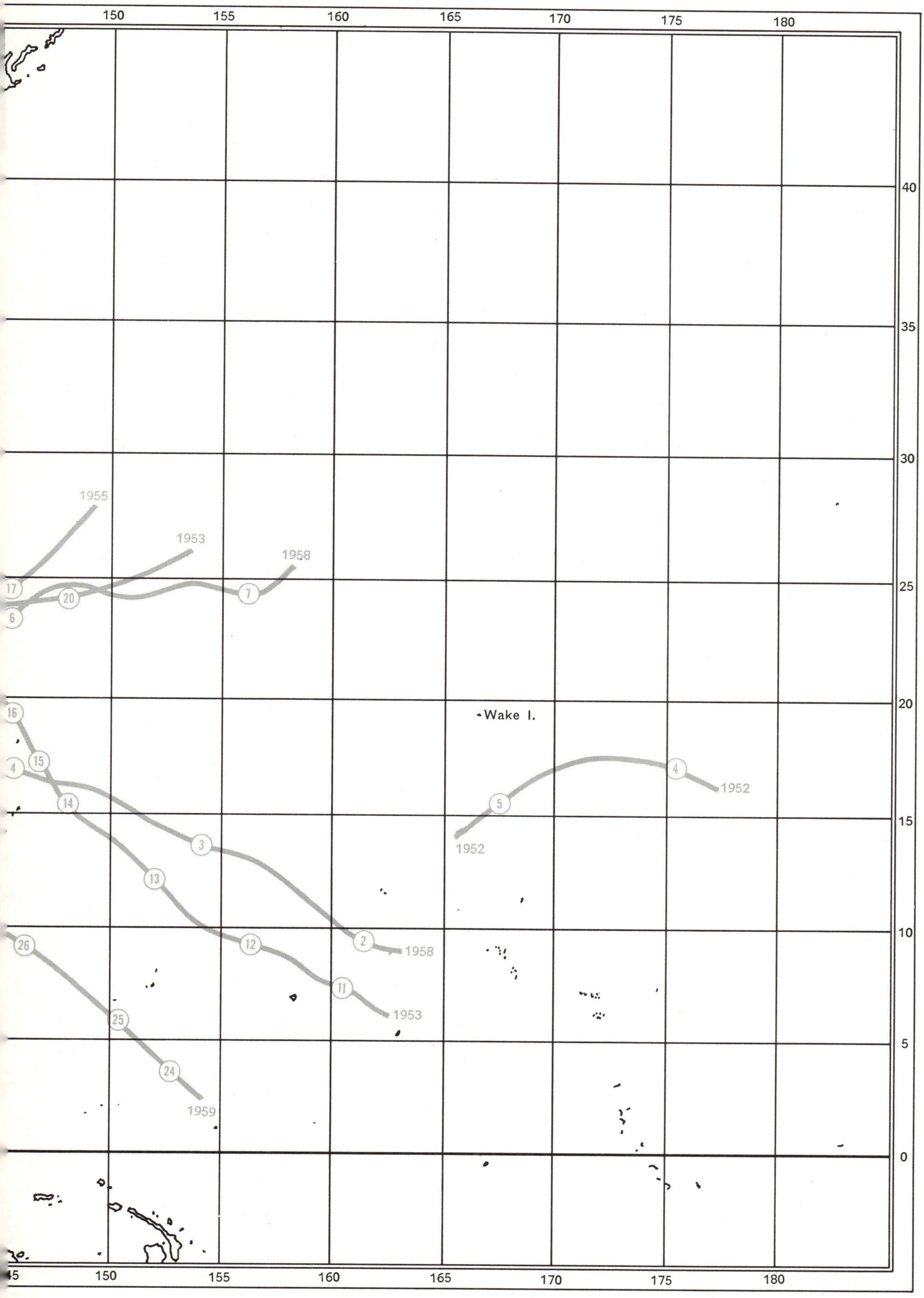
150
155
160
165
170
175
180
40
35
30
25
20
15
10
5
0
1955
1953
1958
17
20
6
7
Wake I.
16
15
4
14
3
13
2
1958
12
11
1953
26
25
24
1959
4
5
1952
1952

Section 3 : Characteristics

All students of typhoons—ships' captains, military commanders, Trust Territory civilian officials, theoretical and operational meteorologists, flight crews, etc. —realize that no two identical typhoons have ever been observed. Nevertheless, there are certain similarities amongst typhoons which do exist and which can be discussed for operational and planning purposes.

Wind

The world's highest typhoon winds have never been scientifically recorded simply because there are no wind instruments in WESTPAC which will withstand a giant typhoon's full fury. Wind instruments now in existence collapse and are carried away as the typhoon approaches—long before the highest intensity of the storm is reached. Current data and estimates are based primarily on: (1) aircraft reconnaissance flights, (2) calculations from pressure-gradient measurements and pressure-temperature relationships, and (3) detailed studies of structural damage to buildings.

Most typhoons over the open ocean areas of WESTPAC have maximum sustained winds of about 140 to 150 knots sometime during their life histories. The distribution of wind speed around an average WESTPAC typhoon over water is shown in Figure 30. This diagram shows a maximum of 150 knots (173 mph) to the right of the direction of movement of the typhoon—looking downstream. It is typical of a moving typhoon that the strongest winds occur to the right. This is due, in part, to the fact that on the right side the forward movement of the typhoon is added to the observed wind velocity. On the left side, the forward movement is subtracted from the observed wind velocity. In the case of a stationary typhoon, the distribution of wind speeds is much more symmetrical.

Certain evidence is available from gust recordings which indicates that the extreme gusts frequently exceed the maximum sustained winds by as much as 30 to 50 per cent. This means that for a 150-knot typhoon, one should prepare for extreme gusts as high as 225 knots.

In the Northern Hemisphere, the winds blow counterclockwise around a typhoon (or any cyclone), as illustrated in Figure 31; in the Southern Hemisphere, the winds blow clockwise around a typhoon (or cyclone). Note in Figure 31 that the winds tend to blow toward the center of a typhoon until they reach the edge of the "eye." The angle of inflow is not the same in every quadrant, but it averages about 20° to 30° across the isobars (lines of constant pressure), except near the inner edge of the wall cloud, where there is no longer any inflow. Wind directions in the diagram have been slightly exaggerated for purposes of illustration.

The angle of inflow varies considerably with individual typhoons, their size, maturity, latitude, and other meteorological

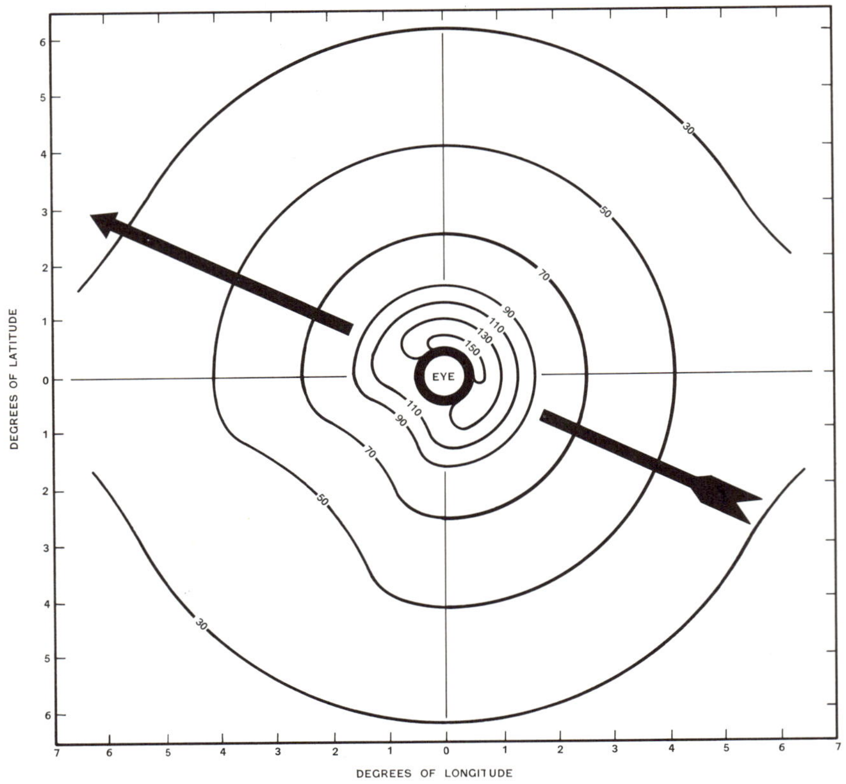

Fig. 30 *Distribution of surface wind speeds (in knots) around an "average" 150-knot typhoon over open water and moving on a track of 295°.*

Fig. 31 *Wind circulation around an "average" 150-knot typhoon in the Northern Hemisphere. Note that in the Northern Hemisphere, the winds blow counterclockwise around the typhoon (or any cyclone). Each pennant indicates 50 knots of wind and each barb indicates 10 knots. Thus, two pennants and two barbs equal 120 knots, and one pennant and two barbs 70 knots.*

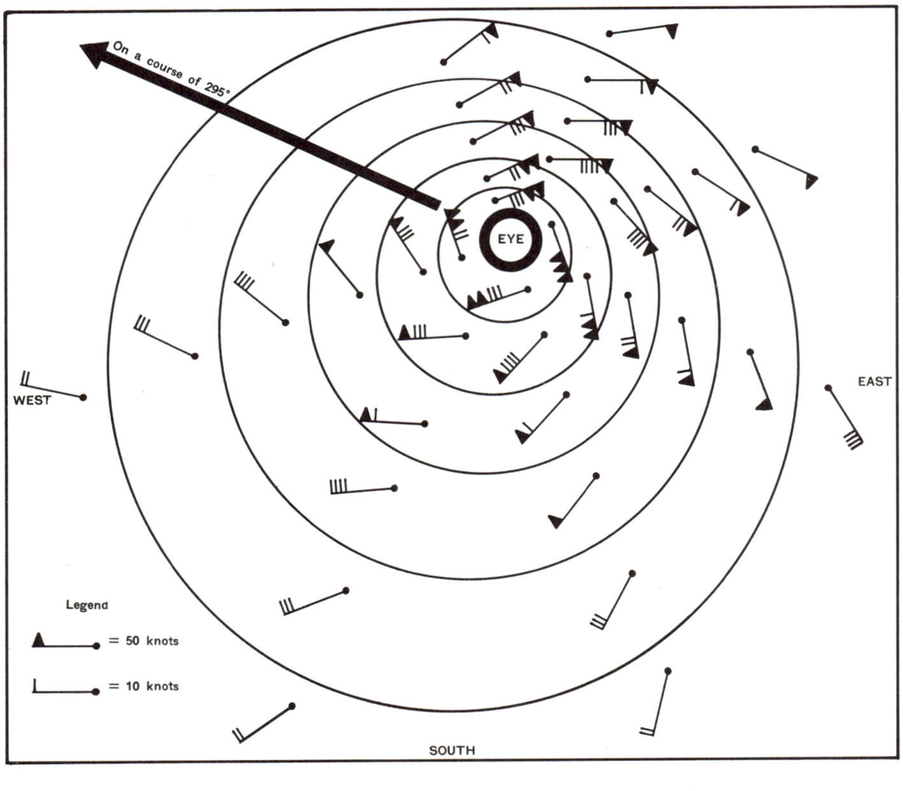

conditions. As a general rule, the intense, immature typhoons have the most symmetrical wind fields, and the strongest winds are located in and under the wall cloud surrounding the eye. Figure 32 shows the distribution of wind speeds on Guam as the eye of Typhoon Karen passed over the island the night of 11-12 November 1962. In poorly developed typhoons, typhoon-force winds are frequently observed in only one quadrant; this is usually the northeast quadrant if the typhoon is moving westward.

Pressure

Motion in the earth's atmosphere, or wind, is caused by the differences in pressure between two points. In general, the winds tend to blow at a certain angle from higher pressure toward lower pressure. And the greater the pressure difference, the faster the wind blows. It is obvious, then, that extremely strong winds, such as those found near the centers of typhoons, can exist only in areas where the pressure differences are extremely large.

The pressure at the center of a typhoon must be very low in order to concentrate a rapid decrease in pressure in a short horizontal distance, resulting in the devastating winds characteristic of the typhoon. Consequently, in a general way, the pressure in the center of a typhoon is a measure of its intensity—although not an absolute measure. To the writer's knowledge, the lowest scientifically accepted sea level barometer reading occurred in a typhoon 460 miles east of the island of Luzon—886.56 millibars (26.18 inches). As a general rule, the lower the pressure in the center, the more intense the typhoon.

Dr. Robert D. Fletcher (Director, Scientific Services, Air Weather Service) developed the following simple formula for calculating the maximum sustained winds of a typhoon:

$$\text{Max. Wind} = 16 \times \sqrt{P_n - P_o},$$

where: P_n = pressure in millibars along the typhoon's periphery.

P_o = pressure in millibars at the center.

If one uses an average value of 1010 millibars for P_n, a very close approximation of the maximum wind is obtained. Taking Typhoon Karen as an example, the pressure in the eye (or center) was 907 millibars. Substituting in the Fletcher formula:

$$\begin{aligned}\text{Max. Wind} &= 16 \times \sqrt{1010 - 907} \\ &= 16 \times \sqrt{103} \\ &= 16 \times 10.1 \\ &= 161.6 \text{ knots}\end{aligned}$$

The actual maximum sustained winds were 150 knots, so the formula provides a rather close answer, slightly on the high side.

Barograph traces at shore stations during a typhoon's passage furnish a measure of the pressure field in the core if the rate of movement is known. Figure 33 is a typical example. The pressure will normally fall extremely rapidly for three hours as the eye approaches, and then rise at an equally rapid rate after passage. Pressure falls of as much as 40 millibars (1.181 inches) in 20 minutes have been recorded, and total pressure drops of 60 millibars (1.772 inches) in 50 miles horizontally are not uncommon. Figure 34 is a tracing of the actual barograph record at the U. S. Fleet Weather Central/Joint Typhoon Warning Center, Guam, as the geometric center of the eye of Typhoon Karen passed nine miles to the south of the instrument's location the night of 11-12 November 1962.

Dr. H. Ito of Japan has completed some outstanding work on the statistical pressure aspects of tropical storms/typhoons and their development. He studied some 259 tropical storms and typhoons which occurred in WESTPAC during the 10-year period 1950-59, inclusive, most of which

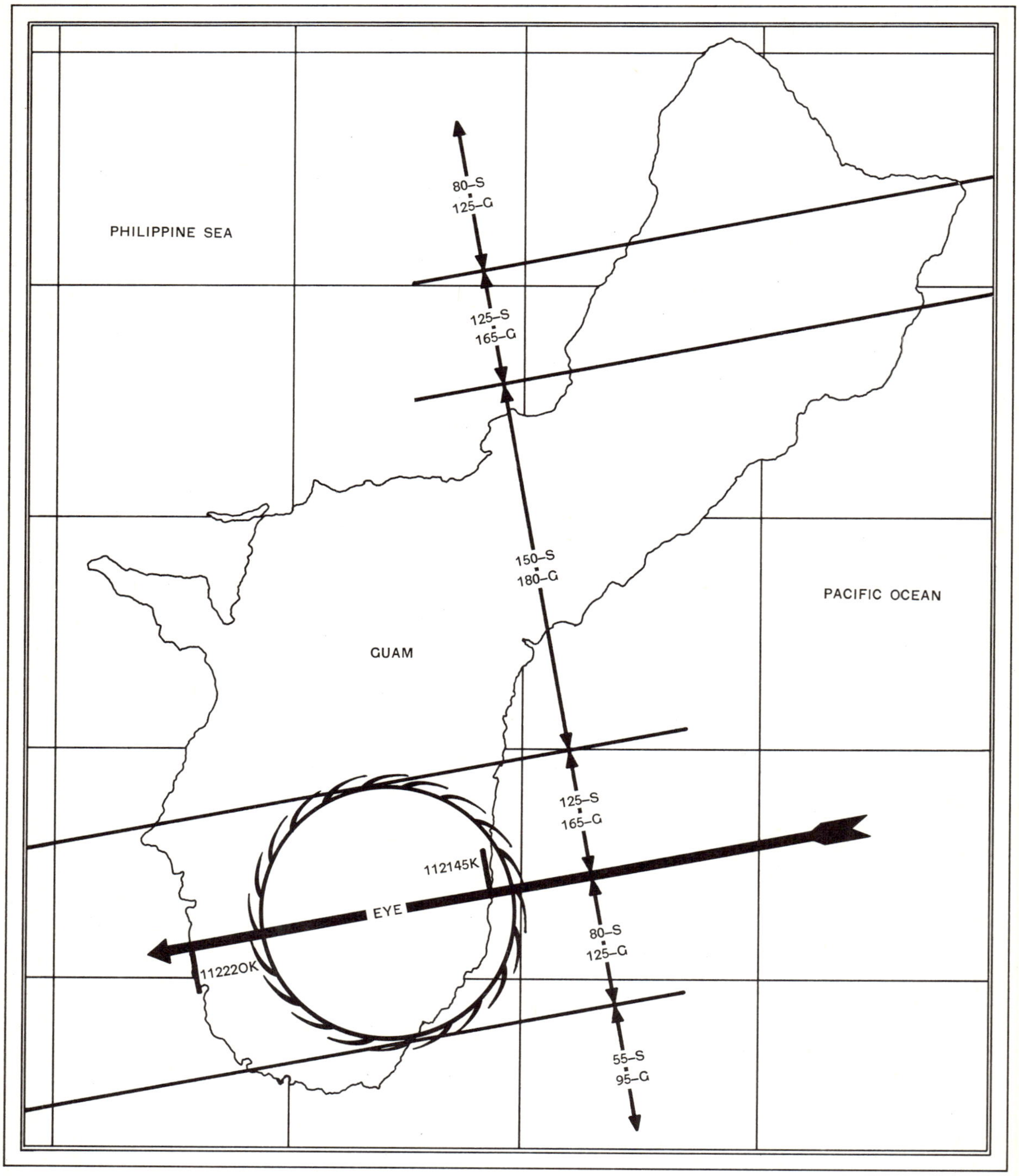

Fig. 32 ***Distribution of wind speeds on Guam as the "eye" of Typhoon Karen passed over the island the night of 11-12 November 1962. S = sustained winds; G = gusts. Even though it took the eye only 35 minutes to cross Guam, the entire island was devastated.***

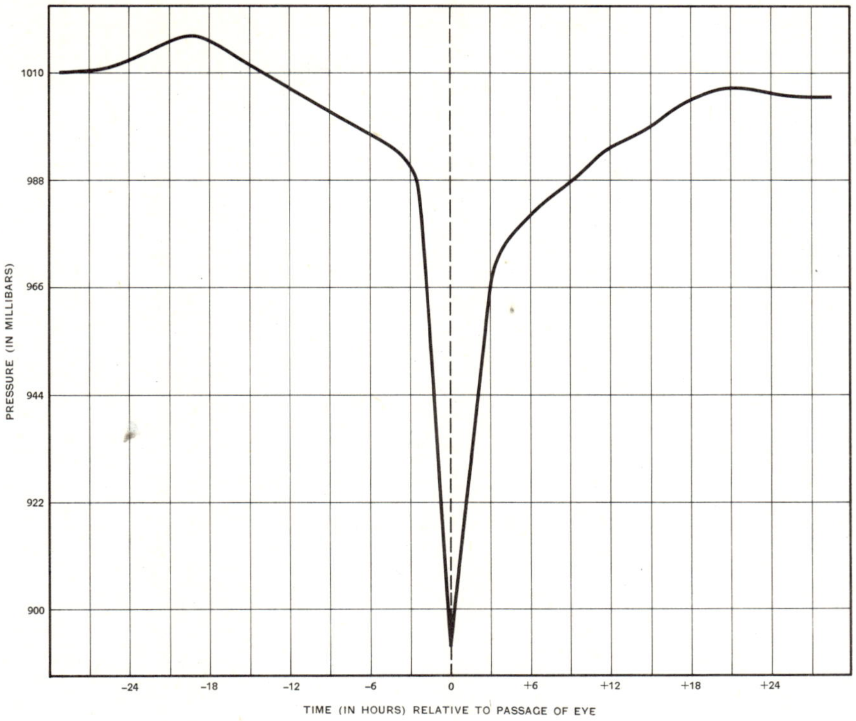

Fig. 33 ***Typical example of a barograph (pressure) trace as the eye of a typhoon approaches and passes over a station or ship.***

Fig. 34 ***A tracing of the actual barograph record at Fleet Weather Center/Joint Typhoon Warning Center (FWC/JTWC), Guam, as the eye of Typhoon Karen passed nine miles to the south of the instrument the night of 11-12 November 1962.***

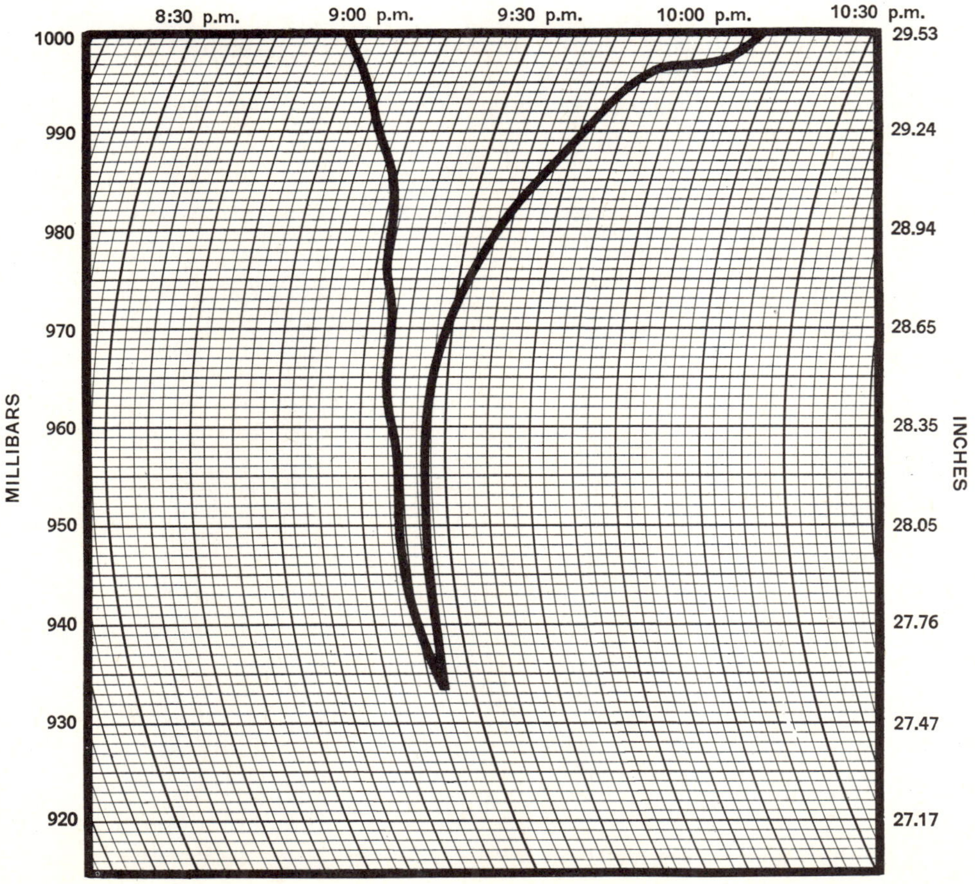

were penetrated by aircraft reconnaissance flights. Figure 35, based on Dr. Ito's research, shows the frequency of occurrence of tropical storms and typhoons, as classified by the lowest surface pressure which each typhoon attained in its life cycle. And Figure 36 illustrates the relation between the lowest central pressure and the maximum 24-hour pressure fall for each typhoon. Dr. Ito also found that, on the average, the central pressures of typhoons begin to drop rapidly about 3 to 4 days after their formation, and that the pressures attain their minimum value about 4 to 5 days after the birth of the typhoon.

Size, Eye, and Vertical Extent

Size. The so-called "size" of a typhoon in WESTPAC is usually expressed in one of three ways, depending upon the particular author: (1) the diameter of the typhoon-force winds (64 knots or higher), (2) the diameter of the 50-knot winds, or (3) the diameter of the outer closed isobar (circular or, most usually, elliptical).

Looking back to Figure 30, it will be found that the average diameter of the typhoon-force winds is about 325 nautical miles. However, in this respect, typhoons are subject to great individual variations. In the case of Typhoon Carmen in August 1960, for example, the typhoon-force winds extended more than 350 miles from the center in all directions, with gale-force winds almost 1800 miles in diameter at one time.

As a rule, in the early stages of development and even for the first day or two after the storm reaches typhoon intensity, the diameter of the typhoon-force winds is usually quite small—on the order of 120 nautical miles. As the typhoon increases in age, it also expands laterally. Sometimes the wind speeds increase markedly, and sometimes they do not, depending upon meteorological conditions both at high levels in the atmosphere and near the earth's surface.

The Eye. The eye of a typhoon is quite unique; it is not observed in any other weather phenomena. Since earliest times, it has been one of the favorite subjects of writers and investigators of tropical storms, for in some respects, it is the most spectacular part of the typhoon.

Rain ceases abruptly at the boundary of a well-developed eye, the sky becomes partly cloudy to mostly clear, and the wind diminishes to less than 12 knots—sometimes to a dead calm.

The average diameter of WESTPAC's typhoon-wind eyes is about 24 nautical miles, although many exceptions to this average figure occur. In immature typhoons, eye diameters of only 4.5 miles have been observed. On the other hand, when the mature Typhoon Carmen passed over Okinawa on 20 August 1960, the diameter of the eye of this monstrous typhoon was observed (and radar-photo verified) to be 200 nautical miles.

Although a typhoon eye is usually described as "circular," it is frequently observed by aircraft reconnaissance crews to be elliptical. When this is the case, the longest axis is usually parallel to the direction of the typhoon's movement; yet, sometimes it rotates. In addition, double eyes and concentric eyes have been observed quite often. Mr. Kenji Shimada noted, in 1961, that the diameter of the eye at 10,000 feet is approximately twice the diameter at the earth's surface. Modern aerial reconnaissance, both visual and radar, has proved that typhoon eyes are constantly undergoing transformation. This is hardly surprising when one considers the extreme conditions at the boundary of the eye.

The length of time it takes for an eye to pass over any point depends upon the size and shape of the eye, the length of the chord which passes over the point, and the speed of movement of the typhoon.

Vertical Extent. Not too many years ago, the vertical extent of typhoons was the subject of greatest controversy amongst theoretical meteorologists. Some insisted that typhoons disappeared above 10,000 feet. Others insisted just as emphatically that they must extend to very great heights. Today—because of high-level

MILLIBARS													
≧1000	999 to 990	989 to 980	979 to 970	969 to 960	959 to 950	949 to 940	939 to 930	929 to 920	919 to 910	909 to 900	899 to 890	889 to 880	879 to 870
14	51	41	24	31	13	14	15	11	11	8	6	3	1
≧29.53	29.50 to 29.24	29.21 to 28.94	28.91 to 28.64	28.62 to 28.35	28.32 to 28.05	28.02 to 27.76	27.73 to 27.46	27.43 to 27.17	27.14 to 26.87	26.84 to 26.58	26.55 to 26.29	26.25 to 25.99	25.96 to 25.69
INCHES													

Fig. 35 Frequency of occurrence of tropical storms and typhoons, classified by the lowest surface pressure which each typhoon attained in its life cycle. (After Dr. H. Ito, 1962.)

Fig. 36 The relation between the lowest central pressure and the maximum 24-hour pressure fall for each typhoon between 1950-59, inclusive. (After Dr. H. Ito, 1962.) Solid curve represents the "mean"; broken line shows the "limits."

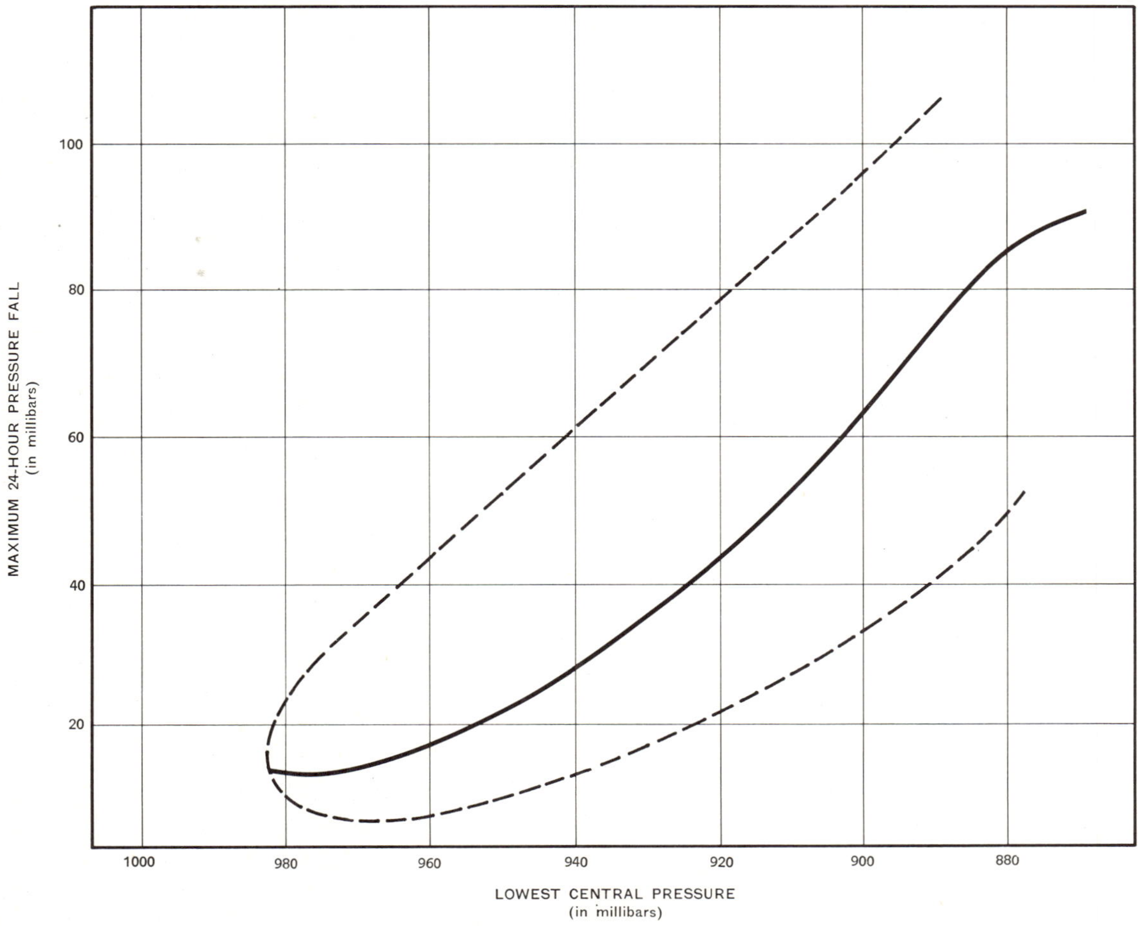

weather observations—the answer is a known fact.

Although typhoons first develop as low-level phenomena, the tops of mature typhoons extend all the way up to the tropopause (the zone between the troposphere and the stratosphere). However, the wind direction reverses completely between the earth's surface and 40,000 feet. Above the low-level cyclonic circulation, there exists an anticyclonic circulation at high levels. Ahead of the center, the winds turn clockwise with increasing height. And to the rear of the center, the winds turn counterclockwise with height. Available data further indicate that the stratospheric easterly winds continue to blow undisturbed, despite nature's rampage taking place underneath.

Clouds and Precipitation

Even though the detailed cloud sequence varies from one typhoon to another, the precursory signs are remarkably uniform. On the day before the typhoon's arrival, local weather is frequently unusually good and the barometer above normal. The arrival of a long swell is sometimes the first warning—coming from the general direction of the storm's center. Then, the pressure begins to fall slowly, and the wind may start blowing from an unusual direction. There are only a few cumulus clouds, and these do not extend very high.

Late in the afternoon—or during the early evening—on the day before the typhoon's arrival, high-level cirrus clouds approach from the direction of the typhoon. These are followed in order several hours later by lowering and thickening cirrostratus, altostratus, and altocumulus clouds. Then, several brief periods of tall cumulus clouds with showers are experienced. The barometer begins to fall much more rapidly, the showers become much heavier, and the wind begins to increase substantially. Finally, a dark wall of clouds approaches—the "bar" of the typhoon. And with its arrival, the full fury of the typhoon is unleashed.

The fury of the typhoon, however, does not persist constantly. Periods of extreme violence alternate with periods of comparative lull. Even the central area of the typhoon is not uniformly filled with clouds. Rather, relatively narrow, elongated bands of very heavy rain are interspersed with areas of light rain—and sometimes no rain at all. Some of these bands have their origin several hundred miles distant from the center of the typhoon. From there, they spiral inward in a counterclockwise direction toward the wall cloud surrounding the eye; hence, the name "spiral bands."

The number of spiral bands around a typhoon varies from a minimum of one to a maximum of eight. The width of each band tends to increase from a minimum of about two miles near the center of the storm to a maximum of 40 miles near the storm's periphery. The general shape of these bands is very close to that of a logarithmic spiral.

Some of the world's heaviest rainfalls have occurred in connection with typhoons. One typhoon, in the year 1911, dumped 88 inches of rain on the Philippines in a four-day period. This is close to the annual average rainfall on Guam.

The rainfall associated with a typhoon is always heavy; however, the exact rainfall is seldom known. After the wind speed exceeds 50 knots, it is unlikely that the rain gauges catch more than 50 per cent, and when the winds go much higher, the rain gauges frequently become a part of the flying debris. The total accumulation of rain at a given locality is, of course, greatly dependent upon the speed of movement of the typhoon, simply because in slowly moving storms the rain lasts a much longer time.

Section 4 : Forces of Destruction

When one considers the fantastic energy aspects of a typhoon, the tremendous destruction wrought by these systems is entirely understandable. For instance, a perfectly ordinary afternoon thunderstorm has the energy equivalent of about thirteen 20-kiloton atomic bombs. And in a 24-hour period, even a small-scale typhoon will release about 20 billion tons of water. This is an energy equivalent of close to 500,000 atomic bombs, or almost six atomic bombs per second.

Wind

Since it is the gustiness of the wind that results in the "uneven" intermittent pressures and the wrenching effects, the speed of peak gusts must be considered in designing ships and structures to withstand the onslaught of typhoons. Another vital consideration in relation to ship and shore construction vis-à-vis typhoon damage is the rapid rise in the actual force of the wind at higher speeds. The force exerted does not increase proportionately with velocity, but rather with the square of the velocity—so that doubling the velocity results in four times the force. Thus, a 52-knot wind results in 15 pounds per square foot, but a 110-knot wind results in the terrific force of 78 pounds per square foot.

The problem is much more complicated, however. Typhoon-force winds passing over and around ships and buildings develop both positive and negative forces. Positive forces are those "pushing in," and negative forces are those "pulling out" (suction effect) externally. And for certain angles of incidence of the wind, the negative pressure may be considerably larger than the frontal force.

Waves and Swell

At the outset, we should distinguish between waves and swell. Waves are generated by the wind, and as these wind-generated waves tend to move out of, or ahead of, the generation area and into regions of weaker winds, they decrease in height and increase in period. These decaying waves, which may persist for a long time, are termed "swells."

The great wind stress exerted on the ocean's surface in typhoons produces huge waves with extreme heights. Some have been reported to be as high as 66 feet. Figure 37 shows schematically how these waves travel outward in all directions from the center of the typhoon. They are observed, as a long swell, as much as 2,000 miles distant from the storm center. The speed of propagation of the long waves is as high as 1,000 miles per day. And since the average typhoon movement prior to recurvature is only 330 miles per day, these swells provide early warning of an impending typhoon.

Referring again to Figure 37, the waves generated in the right rear quadrant of the typhoon travel in the direction of the storm's movement. Thus, they propagate under the influence of winds having rela-

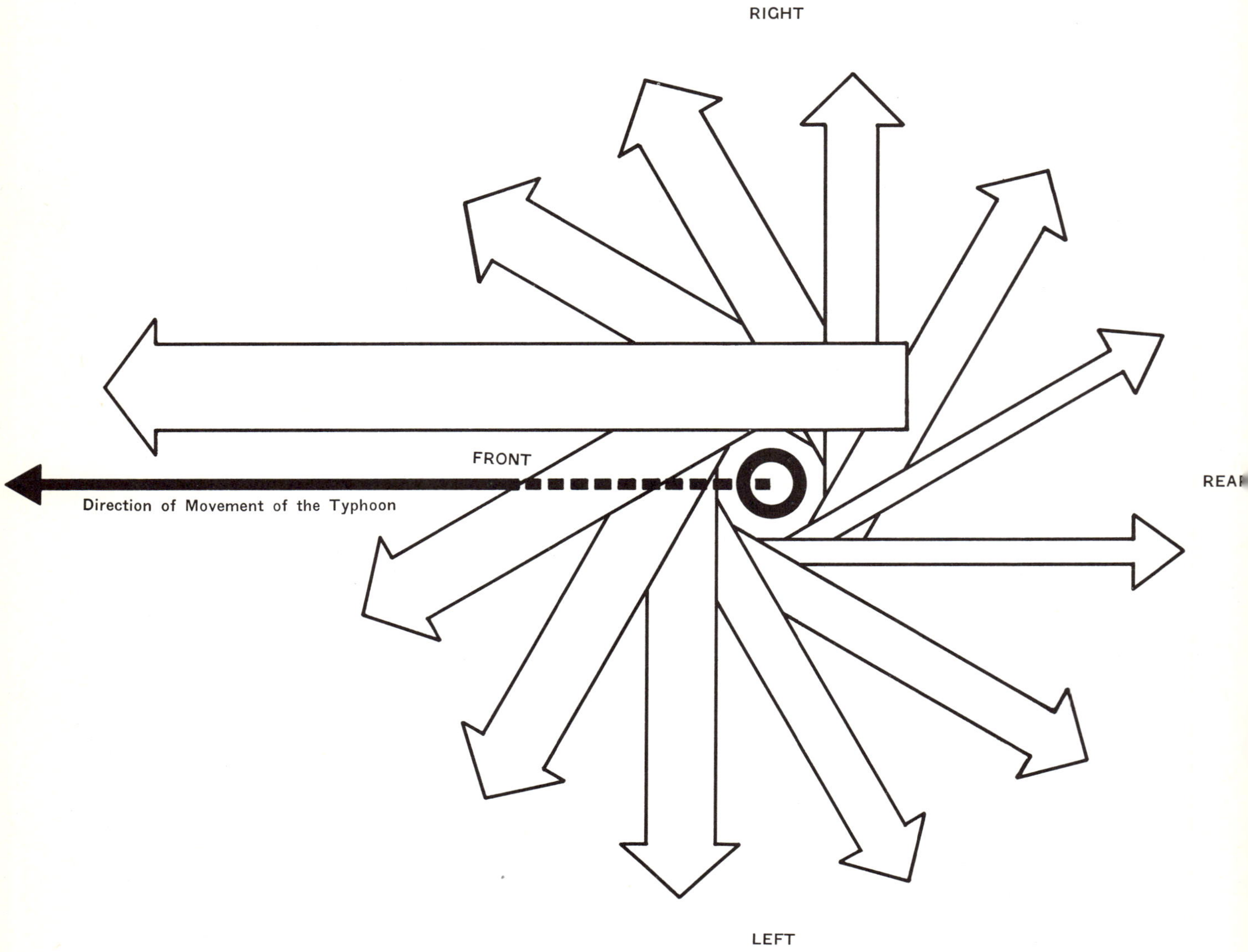

Fig. 37 Schematic diagram of swells generated by a typhoon. Arrows indicate direction of movement of the swells. Width of arrows indicates relative height of the swells.

tively little change in direction for a much longer time than the waves in any other quadrant. Consequently, the highest waves (and later, swells) are produced in this part of the storm. When these swells arrive at a distant ship (or coastline), the normal wave frequency of 10 to 15 per minute will have decreased to about two to five per minute. And the direction of the swell indicates the position of the typhoon's center at the time the swell was generated. If this direction remains constant, the typhoon is approaching the ship (or area) directly. If it changes counterclockwise, as seen by an observer facing the storm, the center of the typhoon will pass from right to left. If the swell changes clockwise, the center of the typhoon will pass from left to right.

One must remember that a cubic yard of water weighs three-fourths of a ton, and a breaking wave moving toward shore at speeds of more than 50 knots is one of the most destructive elements connected with a typhoon. The erosive power of typhoon-driven waves is tremendous, for these waves can scour out 100 feet of beach in just a few hours.

Tides and Surges

A few basic definitions are necessary here to avoid confusion:

Forerunner — The slow rise of the tide produced by swells from a typhoon while it is still a long distance away.

Surge — A rapid rise in the water level produced by onshore typhoon winds and falling barometric pressures.

Water Level — The mean height the water would reach if there were no waves (*i.e.*, the average elevation between wave crests and troughs); the "still water" level.

Meteorological Tides — Departures from normal astronomical tides because of the abnormal action of weather elements.

The forerunner, alone, may produce tides four or five feet above normal, the exact amount depending upon the strength of the wind that produced the swells, the angle between the typhoon's path and the shore line, the shapes of both the coast and continental shelf, etc. It usually causes rises in the water level along several hundred miles of coast, whereas the more rapid and much greater rise associated with the typhoon surge is restricted to a relatively small section of the coastline. The surge normally occurs just to the right of the typhoon's center and a little before or at the same time. When a surge generated on a coast moves into a long channel, however, the rise in the water at the end of the channel may reach its peak several hours after the typhoon winds have passed.

A typhoon moving toward or crossing a coastline will always be accompanied by above-normal tides, particularly near and to the right of the center. The surge may be as little as four feet or as much as 26 feet, depending on several factors. High tides occur at a considerable distance ahead of an advancing typhoon, and here is another warning sign.

There is a very important—though quite rare—sea effect which may occasionally be superimposed upon the usual typhoon tide, resulting in disaster. This has been erroneously called a tidal wave. It has nothing to do with tides, and is usually described as a rapidly advancing "wall of water." Some of the world's greatest natural disasters have occurred as the result of this phenomenon, which, in the writer's opinion, is an unusually rapidly rising and abnormally high typhoon surge.

Floods

Typhoon-produced floods of Japan, Taiwan, Southeast Asia, and the Philippines are likewise no rarity in WESTPAC. Floods have caused the deaths of many thousands of people and damage in the billions of dollars through the years. In general, after sea effects, the major cause of deaths, injuries, and destruction due to typhoons, is floods. And the extent of flooding is dependent upon the physical characteristics of the drainage basin, the rate and total accumulation of rain, and the river stages at the time the rains begin.

Section 5 : Life Cycle and Dissipation

On the average, the life span of WESTPAC's typhoons is about six to seven days from the time they first form until they either enter a land area or recurve into the temperate zone. Some typhoons last for only a day, others a little over two weeks. The evolution of the average typhoon from birth to death is divided into four stages.

Formative Stage

Typhoons form only in pre-existing low latitude disturbances. The fall of barometric pressure can be a slow process, requiring several days for the organization of a large area with diffuse winds. It can also be explosive with a well-formed eye developing in 12 hours. The winds usually remain below 64 knots during the formative stage. The strongest winds are concentrated in one quadrant, usually to the north and east of the center of the developing vortex. The surface pressure drops to about 1002 millibars (29.59 inches).

Immature Stage

All incipient tropical cyclones do not become typhoons. Some have been observed to die within 24 hours, even though the winds had attained typhoon force. Others travel over long distances as shallow tropical depressions (maximum sustained winds not exceeding 33 knots). If intensification takes place, the pressure in the center drops rapidly below 1000 millibars, and winds of typhoon force form a tight band around the center. The cloud and rain pattern changes from disorganized squalls to narrow organized bands spiraling inward toward the center. Up to this point, only a small area is involved.

Mature Stage

The surface pressure at the center no longer falls, and maximum sustained wind speeds have been attained. The lateral area of the circulation expands, and may continue to do so for six or seven days. Whereas winds of typhoon force may blow within a 35- to 50-mile radius during the immature stage, this radius can increase to 350 miles in mature typhoons. However, the symmetry is lost, as the area of gales and bad weather extends much farther to the right of the direction of movement of the typhoon than to the left (looking downstream).

Dissipation Stage

As a typhoon recurves from tropical latitudes to temperate latitudes and enters the belt of westerly winds, its size usually decreases. At times it dissipates within a day. At other times, only its tropical characteristics decay, while the cyclone travels toward the subpolar latitudes, assuming middle-latitude character—and frequently with renewed vigor.

Not all typhoons recurve into temperate

latitudes. Many dissipate over the continent of Asia, and a few over Japan; a small number even die over the tropical Pacific Ocean. The reader is again referred to Figures 16 through 29, which show the tracks of all typhoons in WESTPAC during the 10-year period from 1950 to 1959, inclusive.

Section 6 : Watch Out for Fujiwhara

In the field of hydrodynamics, the interaction between vortex pairs has long been a favorite subject of analysis. Vortices either attract or repel each other. They rotate about a center of gravity located on the straight line—or great circle—connecting them, and the position of this center depends upon the relative mass of the vortices. If they are of equal strength, they rotate about a point situated midway between them. If not, the ratio of the two masses determines the location of the point. Also, in addition to mass, the relative intensity of the circulation within each vortex influences the position of the center of rotation.

The interaction of atmospheric vortices has been studied in great detail by Dr. S. Fujiwhara of Japan; hence, the phrase "Fujiwhara effect." Figures 38 and 39 illustrate this principle. In Figure 38, the northernmost typhoon ("Y") experiences a force acting toward the west from the southernmost typhoon ("X"). In turn, the southernmost typhoon ("X") is subjected to an eastward-directed force. Thus, the axis which may be drawn between the two typhoon centers will rotate counterclockwise with time.

It was precisely this phenomenon which occurred to the south of Japan in late August of 1945 and necessarily delayed urgent diplomatic negotiations.

Since the mid-point itself moves, outside weather systems obviously influence the typhoon pair. If the displacement of the mid-point is subtracted from the tracks, we obtain the relative movement of the two typhoons, and the interaction becomes quite clear. This is shown in Figure 39. The typhoons rotate about each other and are mutually attracted. As they come closer together, their influence on each other grows and the relative rotation increases.

It is the "Fujiwhara effect" which turns the hair of operational meteorologists gray and which forces some ships' captains to become frantic. Yet, in 1962, the captain of a Navy ammunition ship (AE)—en route from Yokosuka, Japan, to Guam—steamed safely through the mid-point between just such a typhoon pair, with no storm damage. However, this is not recommended for general practice.

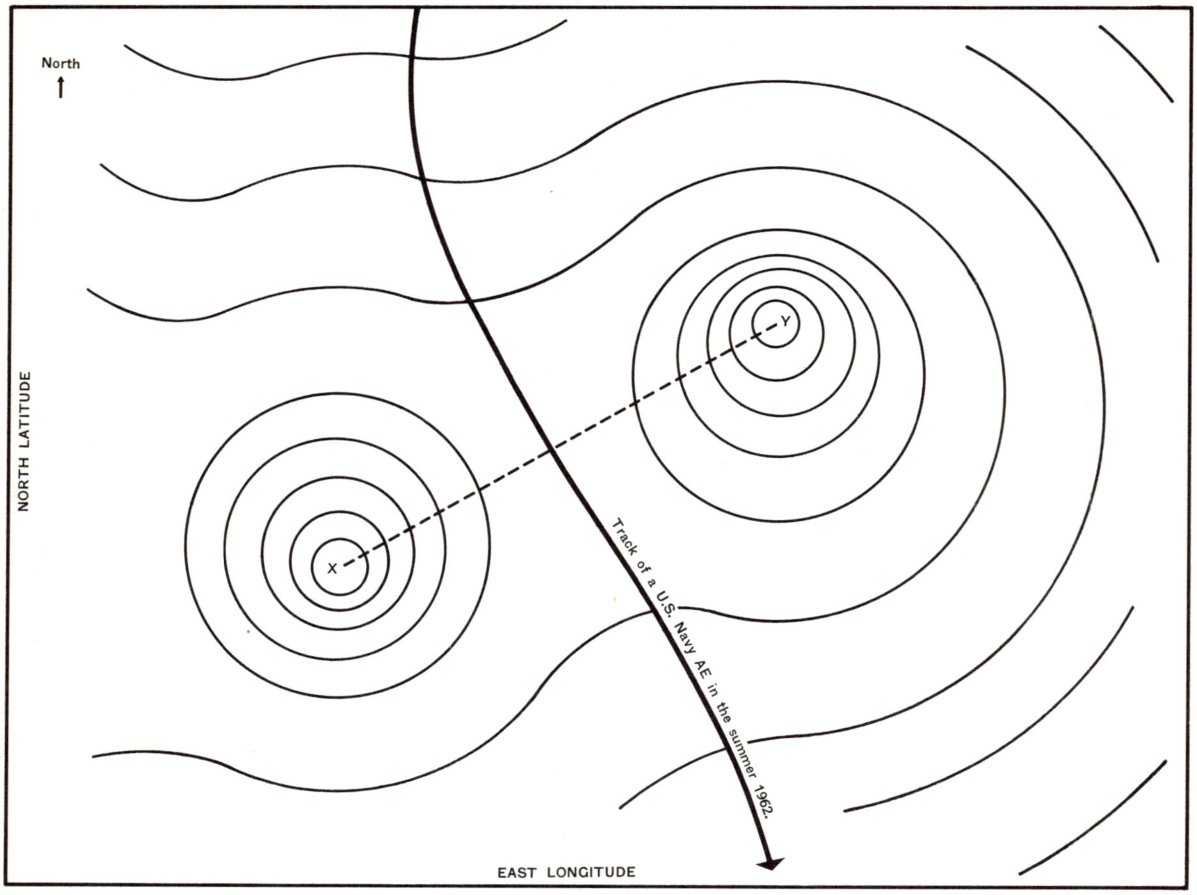

Fig. 38 *A schematic, typical example of a surface weather map with two typhoons which will undergo the "Fujiwhara effect" (i.e., rotating about each other), with typhoon "Y" forced to move toward the west and typhoon "X" forced to move toward the east.*

Fig. 39 *The "Fujiwhara effect." Diagram A shows the tracks of the two typhoons "X" and "Y" and of the mid-point, or center of rotation, for 12-hour intervals. Diagram B shows the displacement of the mid-point subtracted from the tracks, obtaining the relative movement of the two typhoons.*

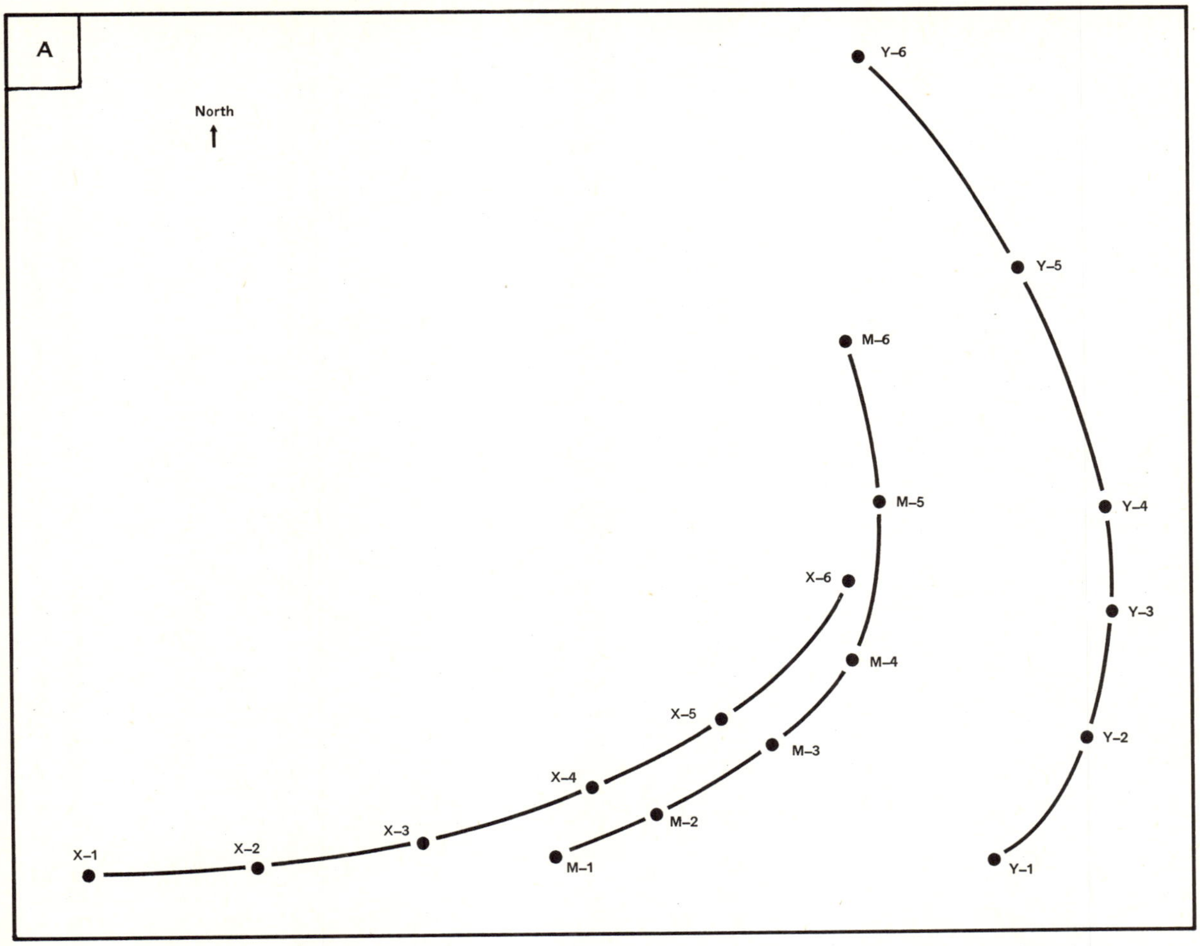
A
North
Y–6
Y–5
M–6
M–5
Y–4
X–6
Y–3
M–4
X–5
M–3
Y–2
X–4
M–2
X–1
X–2
X–3
M–1
Y–1

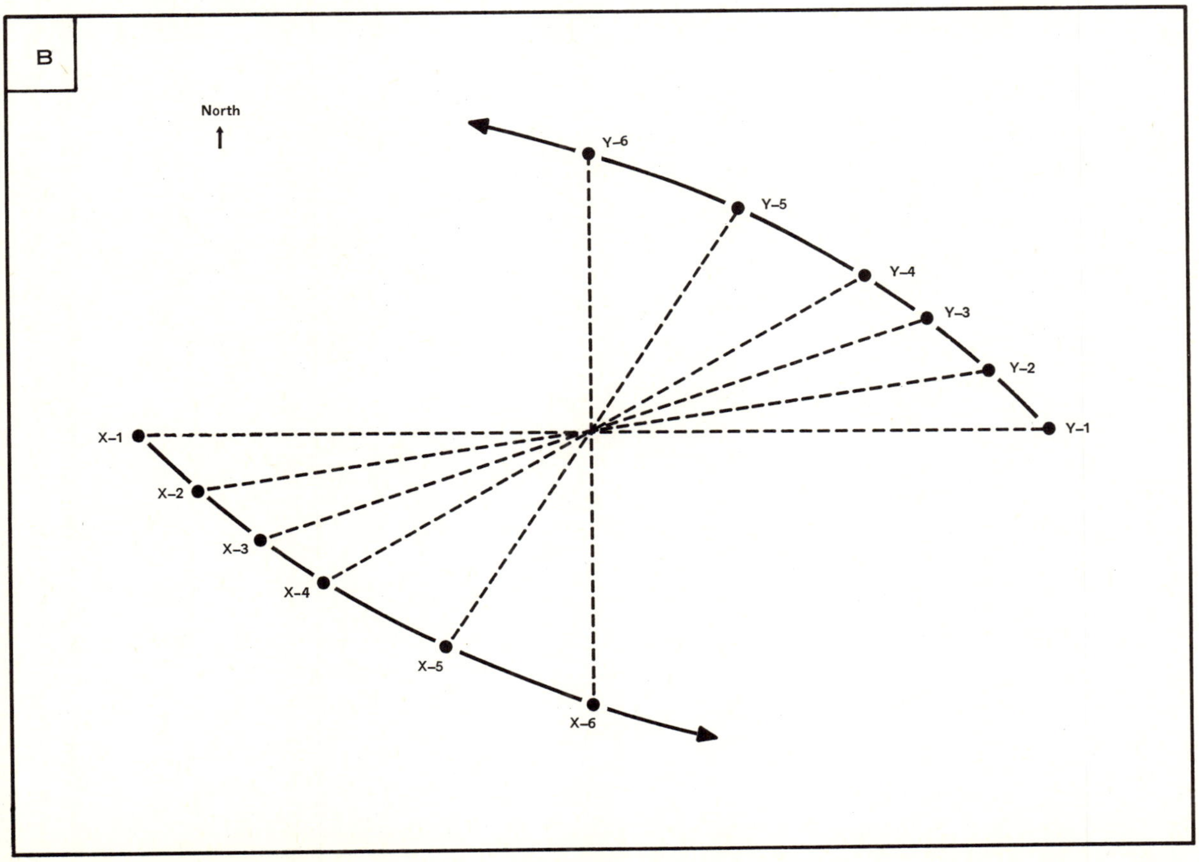
B
North
Y–6
Y–5
Y–4
Y–3
Y–2
X–1
Y–1
X–2
X–3
X–4
X–5
X–6

Section 7 : The Typhoon-Warning System–WESTPAC's "Guardian Angel"

Responsibility for the typhoon-warning system in WESTPAC (west of the International Date Line) is vested in a rather small group of industrious, experienced, and competent operational meteorologists from both the U. S. Navy and the U. S. Air Force. Located atop Nimitz Hill on the island of Guam, overlooking the Pacific, is the nerve center of the system—the U. S. Fleet Weather Central/Joint Typhoon Warning (FWC/JTWC), Guam.

The Joint Typhoon Warning Center (JTWC) department of the over-all FWC/JTWC command has had the following assigned mission since its establishment by Commander in Chief, Pacific (CINCPAC) and Commander in Chief, Pacific Fleet (CINCPACFLT) on 1 May 1959:

1. Provide all tropical cyclone and typhoon warnings to all U. S. Government agencies west of 180 degrees longitude.

2. Determine typhoon reconnaissance requirements and priorities.

3. Conduct investigative and post-analysis programs, including preparation of annual typhoon summaries.

4. Conduct forecast and detection research, as practicable.

The CINCPAC directive further designated the Fuchu Air Force Weather Central, Japan, in co-ordination with the Navy's Fleet Weather Facility, Yokosuka, Japan, as the Alternate WESTPAC JTWC, in the event of FWC/JTWC casualty. This concept has functioned magnificently, and was completely verified as recently as November, 1962.

The efficiency of any system, be it theoretical or operational, is directly proportional to the accuracy and amount of data supplied to the system. Consequently, the great value of each individual surface and upper-air weather and radar report—from land stations, Navy and merchant ships, military and civilian aircraft, etc.—cannot be overemphasized. The typhoon-warning system can only be as good as the data fed into the system. So, be sure to transmit your weather or radar report.

In this regard, because of the vast expanse of the Pacific and the fact that typhoons develop in the low levels of the earth's atmosphere, there is absolutely no substitute for low-level aircraft reconnaissance at or near the 1500-foot level. Admittedly, the fashion and tendency today is to talk about exotic, extremely high-cost systems such as high-speed jets, satellites, solar and lunar probes, and so forth. Nevertheless, the simple truth is that the basic requirement for the typhoon-warning system is a weather-instrumented aerial platform capable of 16 to 18 hours of flight at the 1500-foot level and of penetrating the eye of a typhoon once per day to determine the acceleration and intensification factors.

At the present time, the vital data are obtained by the Navy's squadron AEWRON ONE based at Naval Air Station, Agana, Guam; the Air Force's 54th Weather Reconnaissance Squadron based at Andersen Air Force Base, Guam; and the 56th Weather Reconnaissance Squadron based at Yokota Air Force Base, Japan. However, the Navy's WV-2 aircraft and the Air Force's WB-50 aircraft are getting old and will be phased out of the program in the near future. What type of aircraft will constitute replacements remains to be seen.

Whenever a tropical depression (maximum winds not exceeding 33 knots), a tropical storm (maximum winds 34 to 63 knots), or a typhoon (maximum winds 64 knots or higher) is in existence in WESTPAC, serially numbered warnings bearing an emergency precedence are broadcast every six hours from the FWC/JTWC at 0000Z, 0600Z, 1200Z and 1800Z. These warnings include information relative to the disturbance's location, intensity, and movement, as well as the forecast intensity, direction and speed of movement, and future positions. Formats for tropical depression and tropical storm/typhoon warnings are illustrated in Figures 40 and 41.

All tropical cyclones are assigned a number and it is retained throughout the life of the cyclone. The numbering system begins anew at the start of each calendar year. A tropical depression is assigned the same number that its original cyclone bore. Upon reaching storm or typhoon intensity, the number is replaced by a name taken (in order) from the list contained in Figure 42, progressing down each column before starting on a new column. When the entire list has been used, it is repeated.

Because, primarily, of the present state of the art/science of meteorology and the extreme paucity of weather data in WESTPAC (averaging one report per 226,000 square miles), the average 24-hour forecast error of the JTWC forecasters is about 110 miles, and the 48-hour forecast error is about 240 miles. It must also be remembered that because of the vast expanse of the Pacific, the scarcity of weather data, and the relatively small size of an immature typhoon, it is entirely possible for a tropical storm or typhoon to be in existence for a day or two prior to its discovery by the FWC/JTWC, Guam, and the issuance of appropriate warnings.

In the early stages of the approach of destructive winds from a typhoon, storm, etc., commanders and decision-makers in WESTPAC are responsible for setting the following appropriate weather conditions of readiness:

Condition FOUR	Threat of destructive winds possible within *72* hours.
Condition THREE	Destructive winds possible within *48* hours.
Condition TWO	Destructive winds anticipated within *24* hours.
Condition ONE	Destructive winds anticipated within *12* hours.

Innumerable Navy and merchant marine captains, as well as military and civilian decision-makers, have called upon the FWC/JTWC, Guam, for weather and oceanographic special advice and guidance. Do not hesitate—or be too proud—to do the same. This is a group of dedicated and indefatigable weathermen-oceanographers who will do their very best to meet your requirements.

UNCLAS.

TROPICAL DEPRESSION NINE ZERO WARNING 05.

POSIT ONE THREE PT SEVEN NORTH ONE FOUR ONE PT THREE EAST AT 071200Z. GOOD BASED ON 071000Z RECON FIX.

MAX SUSTAINED WINDS 30 KNOTS.

FORECAST MOVE 310 DEG 12 KNOTS.

FORECAST 081200Z 16 PT 8 NORTH 137 PT 6 EAST. MAX SUSTAINED WINDS 45 KNOTS.

TROPICAL DEPRESSION NINE ZERO RECON FIXES SCHEDULED FOR 072200Z 080400Z AND 081600Z.

BT.

Fig. 40 Tropical depression warning example.

Fig. 41 Tropical storm-typhoon warning example.

UNCLAS.

TYPHOON KAREN WARNING 14.

POSIT ONE THREE PT ONE NORTH ONE FOUR EIGHT PT TWO EAST AT 110000Z. GOOD BASED ON 102200Z RECON FIX.

MAX SUSTAINED WIND 150 KNOTS NEAR CENTER.

RADIUS 100 KNOT WINDS 30 MILES.

RADIUS 50 KNOT WINDS 75 MILES SOUTHWEST SEMICIRCLE AND 150 MILES NORTHEAST SEMICIRCLE.

RADIUS 30 KNOT WINDS 250 MILES SOUTHWEST SEMICIRCLE AND 450 MILES NORTHEAST SEMICIRCLE.

FORECAST MOVE 285 DEG 15 KNOTS FIRST 12 HOURS THEN 280 DEG 17 KNOTS NEXT 12 HOURS.

FORECAST POSIT 111200Z 13 PT 8 NORTH 145 PT 1 EAST MAX SUSTAINED WINDS 155 KNOTS. 120000Z 14 PT 5 NORTH.

141 PT 7 EAST MAX SUSTAINED WINDS 155 KNOTS. OUTLOOK 130000Z POSIT 14 PT 9 NORTH 134 PT 7 EAST.

KAREN FIXES SCHEDULED FOR 110400Z, 111000Z AND 111600Z.

BT.

NAMES FOR TYPHOONS IN WESTPAC			
1	2	3	4
ALICE	ANITA	AMY	AGNES
BETTY	BILLIE	BABE	BESS
CORA	CLARA	CARLA	CARMEN
DORIS	DOT	DINAH	DELLA
ELSIE	ELLEN	EMMA	ELAINE
FLOSSIE	FRAN	FREDA	FAYE
GRACE	GEORGIA	GILDA	GLORIA
HELEN	HOPE	HARRIET	HESTER
IDA	IRIS	IVY	IRMA
JUNE	JOAN	JEAN	JUDY
KATHY	KATE	KIM	KIT
LORNA	LOUISE	LUCY	LOLA
MARIE	MARGE	MARY	MAMIE
NANCY	NORA	NADINE	NINA
OLGA	OPEL	OLIVE	ORA
PAMELA	PATSY	POLLY	PHYLLIS
RUBY	RUTH	ROSE	RITA
SALLY	SARAH	SHIRLEY	SUSAN
TILDA	THELMA	TRIX	TESS
VIOLET	VERA	VIRGINIA	VIOLA
WILDA	WANDA	WENDY	WINNIE

Fig. 42 List of WESTPAC typhoon names.

Section 8 : Typhoon Tips from a Weatherman

Experience has clearly shown that for storm/typhoon forecasts and warnings to be highly effective, they must be: (1) issued at frequent intervals, (2) issued at regular times, and (3) worded in specific terms. This is precisely the case in WESTPAC; yet despite this fact, individuals and units, etc., still get into trouble.

There are many reasons for this: small immature typhoons can be in existence unbeknownst to the FWC/JTWC; the 12-, 24-, and 48-hour forecasts of future positions and intensities are not 100 per cent accurate; no two individuals in positions of command and no two decision-makers think or act exactly alike; and so forth.

The position of command or authority is where the decision is made. Advice and/or guidance should be sought from the FWC/JTWC; however, the final decision rests with the Commanding Officer—or other decision-maker—as to which course of action to pursue. This was well expressed by Commander, Task Force 73 in a message in August, 1962:

> . . . THE COMMANDING OFFICER IS IN BEST POSITION TO DETERMINE OWN COURSE OF ACTION BASED ON LATEST INFORMATION . . . .

There are, however, a few "typhoon tips" which the writer of this guide advances for consideration by WESTPAC sailors and decision-makers.

Afloat and Ashore

1. Read through this guide at least twice at your leisure, and keep it in a convenient place for quick reference.

2. Know and keep handy your Office of Chief of Naval Operations (OPNAV) and CINCPACFLT INSTRUCTIONS in the 3140.*xx* series. These deal with the general subjects of weather and oceanography, and the numbers and types of reports required from you. In particular, be familiar with CINCPACFLT INSTRUCTION 3140.6*x*, which deals with the joint typhoon warning service and procedures in WESTPAC.

3. If the surface pressure drops to 1002 millibars (29.59 inches) in your location, start thinking about preliminary buttoning-up operations, whether you have received a warning from the FWC/JTWC or not. If the surface pressure drops below 1000 millibars (29.53 inches), start preliminary buttoning-up. The receipt of your weather report to FWC/JTWC may be (or may have been) delayed for some reason.

4. If you notice an unusual long swell with a frequency of about 2 to 5 per minute, the chances are that there is a storm or typhoon at some distance from your location in the general direction from which the swell is approaching.

5. If the typhoon is anywhere near your position (location), keep a detailed plot of the 6-hourly warnings broadcast by the FWC/JTWC.

6. If the eye passes over your ship or location, do not leave your place of safety. Shortly after the calm, the wind will begin to blow even more furiously from the opposite direction.

7. Remember that the average 24-hour forecast error is about 110 miles and that the average 48-hour forecast error is about 240 miles. Make these allowances for all warnings received.

Afloat

1. Never try to outrun, or pass ahead of, a typhoon if there is some other course of action available to you. Chances are you will not make it. Remember that the typhoon-generated swells move ahead of the storm at speeds close to 50 knots. Once these swells arrive at your ship, your Speed of Advance (SOA) is cut down to 4 to 5 knots. You are a sitting duck and the eye could pass right over you.

2. The winds are the best indication of the direction of the center of the typhoon from your ship. Memorize Figure 31. Adding 115° to the direction from which your observed true wind is blowing gives the approximate bearing from the ship to the typhoon's center. If the ship bears 090° from the center, the observed true wind would be from 155°. Adding 115° to 155° = 270°, the bearing of the eye from the ship. For sums higher than 360, subtract the number "360" to obtain the bearing.

3. If circumstances preclude avoiding the typhoon altogether, the proper action to take depends upon the ship's position relative to the typhoon's center and the direction of travel of the typhoon. If possible, avoid the dangerous semicircle (the right semicircle, looking downstream, in the Northern Hemisphere). Try to maneuver into the navigable semicircle (the left semicircle, looking downstream, in the Northern Hemisphere). The reverse is true in the Southern Hemisphere.

4. A gradually veering (changing clockwise) wind indicates a position in the dangerous semicircle. A backing (changing counterclockwise) wind indicates a position in the navigable semicircle.

5. If the wind direction remains steady with increasing speed and rapidly falling pressure, the ship is near (or in) the path of the typhoon. If the wind direction remains steady with decreasing speed and rising pressure, the ship is on the typhoon track, but in a safe position behind the center.

6. Keep that radar "fired up." This will give you bearing and distance to the eye, as long as the equipment holds up.

7. Do not guess. Work out your relative movement plot and Closest Point of Approach (CPA) on a maneuvering board or radar scope.

8. In general, if unavoidably caught within the circulation of a typhoon—other things being equal—the best courses of action are as follows:

a. *In the Right or Dangerous Semicircle*
Bring the wind on the starboard bow (045° relative), hold course and make as much speed as possible.

b. *In the Left or Navigable Semicircle*
Bring the wind on the starboard quarter (130° relative), hold course and make as much speed as possible.

c. *On the Storm Track—Ahead of the Center*
Bring the wind two points on the starboard quarter (158° relative), hold course and make as much speed as possible. When well within the navigable semicircle, bring the wind to 130° relative.

d. *On the Storm Track—Behind the Center*
Steer the best riding course which maintains opening with the ty-

phoon center. Remember that typhoons tend to curve northward and recurve eastward.

Ashore

1. Sterilize bathtub, bottles, jugs, cooking utensils, etc., and fill with water prior to typhoon's arrival, for water will probably become contaminated.

2. Raze weak structures; bring garbage cans, garden tools, porch and lanai furniture, awnings, TV antennas, etc., inside. Loose objects become veritable missiles of death and destruction during a typhoon.

3. Have a good supply of canned food, charcoal briquets, candles, flashlights and extra batteries and bulbs, matches, tobacco, a bottle of your favorite spirits, and a first-aid kit.

4. Keep a few carpenter's tools (hammer, saw, pliers, small crow-bar, nails, etc.) and some "boarding-up" lumber handy for emergency repairs.

5. Keep a battery-powered radio turned on at all times for latest official information. Do not be misled by rumors.

6. Evacuate and avoid low-lying coastal areas which may be swept by tides, waves, or surges.

7. Board up windows, but be sure that a window or door can be opened on the lee side of the structure to keep pressures on inside and outside of house equalized.

8. Turn refrigerator and freezer to the coldest setting. Open only when absolutely necessary, and close quickly. Both will remain colder much longer. These will maintain food-preserving temperatures 48 to 56 hours, and then dry ice can be used.

9. Fill the gas tank of your car prior to storm's arrival.

10. Disconnect main electrical switch to prevent short circuits when current is restored, and shut off gas valve until after post-typhoon examination.

11. Remain calm and cautious; panic spreads rapidly.

Section 9 : Outlook for the Future

If there is the impression that a great deal must still be learned about typhoons by the theoretical and operational meteorologists, that impression is absolutely valid. There are still many unanswered questions about typhoons, but progress is being made every day. For example, very few typhoons engulf or strike a ship or geographical area without prior adequate warning. The greatest difficulties involve precision forecasts of the exact time of formation, the maximum intensification, the exact height of storm tides and surges, and the exact position 48 and 72 hours in advance.

The U. S. Weather Bureau, the U. S. Navy, and the U. S. Air Force are all pressing forward in co-ordinated, accelerated research programs embracing all facets of typhoons and hurricanes, and progress has been steady. The National Hurricane Center—constituted in the spring of 1959—has been a beehive of activity.

Individual weather and radar reports, and aircraft reconnaissance, are the backbone of the typhoon-warning system in WESTPAC. Without these individual and vital reports and services, the entire system would collapse. The great importance of each ship and aircraft report (both military and civilian) cannot be overemphasized.

Satellites now contribute—and will continue to contribute—to the typhoon-warning service. However, since incipient typhoons are shallow phenomena, the basic and vital requirement is for low-level meteorological data. And these are obtained by land stations, ships, and weather-instrumented aircraft capable of sustained flight near the 1500-foot level. To be sure, our current (and future) satellite systems do (and will) afford a contribution to the typhoon-warning service, but the real answer lies with the old-fashioned WB-50 and WV-2 aircraft—or newer aircraft with similar capabilities. Unfortunately, these older aircraft are now being retired.

Extremely long-range radars (now under design) are only a partial answer. If one is willing to accept a system of merely "tracking and extrapolating" typhoons, radar alone will provide the answers. However, radar will not indicate whether the typhoon contains 80-knot or 180-knot winds, nor will it provide acceleration or intensification factors for future forecasts.

Electronic computers are another new approach to the problem, and their results are also being currently utilized by the FWC/JTWC, Guam. However, the products of these "machines" are only as good as the data fed into them and the individual programming. And we are right back where we started—the acquisition of low-level weather data.

It has been suggested—and considered—that incipient typhoons be seeded with dry ice, silver iodide, or carbon black, for the purpose of either dissipating or diverting them. This remains a very controversial subject amongst theoretical and operational meteorologists and political decision-makers. If typhoon control proved eminently successful and if no storms were permitted to develop and to realize a full life cycle, would not nature find some other method of maintaining the earth's heat balance? Perhaps this would be even more disastrous than the typhoon itself.

CAPTAIN WILLIAM J. KOTSCH, U. S. NAVY

In 1941, after graduating from Tusculum College in Greeneville, Tennessee, where he majored in mathematics and chemistry, Captain Kotsch undertook graduate study in meteorology at the Massachusetts Institute of Technology. He was commissioned as an officer in the U. S. Navy in 1942.

During World War II, he served as Aerological Officer of ACORN # 33 and U. S. Naval Air Base # 943, and as Executive Officer of the Fleet Weather Central, Guam.

Captain Kotsch's subsequent assignments have been: Assistant Head of Department, Aerological Engineering, U. S. Naval Postgraduate School; Staff Meteorologist, Commander Second Fleet; Assistant Officer in Charge, Project AROWA (Applied Research and Operational Weather Analysis); Chief Forecaster for the first H-bomb detonation at Eniwetok in November 1952; Office of the Chief of Naval Operations, Naval Weather Service Division; Officer in Charge, Fleet Weather Facility, London; Student in Senior Course, Naval War College; Deputy Director, Strategy Study Division, Naval War College; Commanding Officer, Fleet Weather Central/Joint Typhoon Warning Center, Guam, where his command was almost destroyed by Typhoon Karen in November 1962.

In 1961, Captain Kotsch obtained his master's degree in political science at Boston University. He is presently the Director, Meteorology/Oceanography on the staffs of CINCLANT, CINCLANTFLT, and CINCWESTLANT.

APPENDIXES

Appendix I : 400 Years of Hurricanes*

By Chief Aerographer's Mate Charles R. Coates, U. S. Navy

Hurricanes, like other weather phenomena, have been in existence since the beginning of time. But on his first trips to the Caribbean, Columbus identified the tropical storms there as being different in some way from the extratropical disturbances of the Mediterranean. On his last trip to the New World, Columbus encountered a severe hurricane which brought to the attention of all concerned the major characteristics of these tropical storms. This storm created widespread damage to the Spanish fleet with a loss of 32 ships and hundreds of lives.

From the time of the bad storm in 1502 until the present day, widespread interest has arisen concerning the action, characteristics, and peculiarities of the tropical storms which occur each year at some location along the Gulf of Mexico, Caribbean, or the Atlantic. Statistics vary as to the average number which may occur in any one year. As many as 20 storms have been tracked within one season from June and through November. During some years there has been no storm of significance.

In the 456 years since Columbus came to the New World, the history of the Caribbean has been punctuated by the action of the tropical storm—the hurricane. Even the name is native, being derived from the Carib god Huracan whose frown makes livid the sky and whose lips, pursed for blowing, create the circular tempest. No resident or frequent traveler to the Caribbean can long count on missing the phenomenon of the hurricane in all its infamous activity.

Santo Domingo, the capital of the Spanish in the island of Española, appeared to be extremely susceptible to hurricanes of the early period. On August 3, 1508, a severe hurricane visited the city with heavy damage. Again in July, 1509, almost the entire city of Santo Domingo was destroyed by a hurricane. Over half of the observed hurricanes of the sixteenth century were reported in the Santo Domingo vicinity.

A review of the hurricane characteristics will show why the number reported in the sixteenth century was small, and why Santo Domingo seemed to be the center of activity. The storms originate in the doldrums of the Atlantic, a little north of the equator, and are caught up by the easterly trades to be sent spinning into the Caribbean. But because hurricane winds seldom occur more than fifty to one hundred miles from the storm center, many storms pass unnoticed in sparsely settled areas. Often the heavy rainfall on the outer rim of the hurricane is a godsend to the dried up vegetation, and the benefits far outweigh the squally weather over the area.

A true interest in the study of the hurricane had to wait until the Caribbean was more densely settled, about 1550. Roberts,

* From United States Naval Institute *Proceedings*—May 1949.

in his book *The Caribbean,* reports in detail the extent of colonization in the year 1542. The report shows the growth of the area during the first fifty years, but what is more important to the student of weather is that it shows the spread of population who may report the occurrence of severe tropical storms.

The census of 1542 shows Española as still the center of cultural and social interests, with the largest part of the Spanish colonial population living here and along the eastern coast of Cuba. Puerto Rico had been settled, but little had been done in colonizing the Lesser Antilles extending from Antigua to Trinidad. Most of the population in South America settled along the coast, with the two centers of population at Panama and Cartagena. Parts of Central America were settled, with the largest colony at Vera Cruz. Florida had only scattered settlements of an intermittent nature, St. Augustine being the more permanent of the group.

Tracing the average path of the hurricanes on the population map even now, it is easy to see that a large number of the storms will still pass outside the populated centers. But in contrast, travel in the area has grown to the point where many of the storms are reported by transients within the area. One such storm which might have influenced history considerably was that encountered by Sir John Hawkins off Dry Tortugas in August, 1568. This severe storm drove the British fleet ahead of it and into the Spanish port of Vera Cruz where Hawkins later seized San Juan de Ulua fort to dominate the harbor.

History shows that the destiny of St. Augustine might have been entirely influenced by the hurricane which moved over the city in 1565. On September 16, the fate of St. Augustine was secured for the Spaniards when a hurricane struck both the French and Spanish fleets; but because most of the Spaniards were ashore, it was to their advantage, even though all of their heavy ships were lost in the blow. This storm, like many of the other hurricanes, came with widespread harbingers of bad weather if only Ribault, the French commander, had been able to recognize the signs of the weather lore.

No scientific study explained in detail the action of the hurricane until the nineteenth century, but many of the early mariners must have been able to correlate the changes in the sky and oppressive weather with the coming of the hurricane. In 1502 Columbus was vindicated as a prophet because he warned the city of Santo Domingo of the coming storm. Even today the prediction of a storm occurring in this part of the Caribbean during the month of June would be called unseasonable, and draw attention to the persons forecasting such a phenomenon.

Probably the earliest extended and accurate description of a tropical storm is found in the narrative of Alvar Nuñez, treasurer and "High Sheriff" of the expedition of Pánfilo de Narváez, which reached the West Indies in the autumn of 1527. Nuñez had been despatched to the port of Trinidad with two ships to take on supplies. There he describes the weather as looking "ominous," and from this beginning proceeds to log in detail the events of the storm with more than a lay knowledge of its action. Each report in part brings out some of the major differences in tropical and extratropical storms.

Through the eighteenth century the library of hurricane reports continued to grow. The descriptive terms were in general the same, and except for the number of lives lost or the location of the storm they were all similar. Many of the storms had very great influence on the social, educational, and military matters of the day. A hurricane period was established in part, for it was noted early in the history of hurricane study that the worst storms occurred during the month of September. The current folk rhyme of the British West

Indies summarizes the knowledge available to the population of the Caribbean as early as 1565.

June, too soon
July, stand by
August, don't trust
September, remember
October, all over.

The year 1780 is one that stands out in bold relief from other years in the study of the hurricane. This was the year of the "Great Hurricane," thought to be the most violent of the eighteenth century. Colonel Reid says: "Three great storms occurred nearly at the same time; and these have been confounded together, and considered but as one. The first destroyed the town of Savanna-la-Mar, on October 3, 1780. The second, by far the greater one, passed over Barbados on the 10th and 11th of the same month and year. The third dispersed and disabled the Spanish fleet, under Solano, in the Gulf of Mexico, after it had sailed from Havanah to attack Pensacola."

A report on the "Great Hurricane" describes in detail what might be expected in any great tropical storm. The following account by Elisee Reclus, published in 1874, is given here to review that thought. It is only through the forecasts of impending danger that protection is offered the coastal cities of today.

> The most terrible cyclone of modern times is probably that of the 10th of October, 1780, which has been specially named the "great hurricane." Starting from Barbados, where neither trees nor dwellings were left standing, it caused an English fleet anchored off St. Lucia to disappear, and completely ravaged this island, where 6,000 persons were crushed in the ruins. After this, the whirlwinds tended toward Martinique, enveloped a convoy of French transports, and sunk more than 40 ships carrying 4,000 soldiers; on land the towns of St. Pierre and other places were completely razed by the wind, and 9,000 persons perished there. More to the north, Dominique, St. Eustatius, St. Vincent, and Porto Rico were likewise devastated, and most of the vessels which were in the path of the cyclone foundered with all their crews. Beyond Porto Rico the tempest bent to the northeast, toward the Bermudas, and though its violence had gradually diminished, it sank several English warships returning to Europe. At Barbados, where the cyclone had commenced its terrible spiral, the wind was unchained with such fury that the inhabitants hiding in their cellars did not hear their houses falling above their heads; they did not feel the shocks of earthquake which, according to Rodney, accompanied the storm.

The description of this storm was in terms of scientific knowledge accumulated from extended study during the last part of the eighteenth and first part of the nineteenth centuries. Use of the word "cyclone" and the movement of the storm along a track was part of the modern knowledge coming into use at this time. Piddington proposed the name "cyclone" to express the "tendency to circular motion." In his *Sailor's Hornbook,* Mr. Piddington credited Captain Langford with the earliest published statement of the nature of tropical storms appearing in the publication *Philosophical Transactions* in 1698. But nearly fifty years before, a German geographer, Bernhardus Varenious, treated hurricanes as whirlwinds in a book entitled *Geographia Naturalis.*

William Dampier, a master seaman and world voyager, wrote much in detail about the winds within the tropical storms. He describes conditions experienced in a typhoon off the coast of China in the year 1687, at which time he came to the conclusion that there was no difference between the typhoon and the hurricane except in name.

In 1819 Professor H. W. Brades, of the University of Breslau, worked out the idea of the synoptic weather map. Even though he had to use reports that were old and of no value for forecasting, he was able to set up a detailed study of the laws of the storms. From observations taken in 1783 Brades established the circulation of winds in a cyclone. The study was picked up by Heinrich Wilhelm Dove, physicist and meteorologist, to show that cyclones rotate clockwise in the southern hemisphere as compared to counterclockwise in the northern hemisphere.

William C. Redfield was enabled to give a more fundamental account of the phenomena of rotation and translation based upon observations of West Indian hurricanes collected over a ten-year period. In 1831 he defined a cyclone as consisting of a large mass of air with a rapid motion of rotation counterclockwise, a calm center, and a motion of translation. By sequent researches he identified the region between the equator and the Tropic of Cancer as the place of origin of the storms. The scientific study of cyclones and hurricanes was well underway by the middle of the nineteenth century, but still no use of the knowledge had been made in trying to forecast the coming of the storms.

Following the invention of the weather map by Brades, some system of relaying current weather reports had to be improvised before accurate forecasts could be issued on a daily basis. Collection of weather observations was greatly facilitated with the invention of the telegraph in 1843. Ten years later an attempt was made to collect daily weather reports from a network of stations for the sole purpose of issuing weather forecasts to the public.

The science of forecasting had been well developed long before it was possible to collect information from ships at sea. Most of the Caribbean network of stations had to wait until the beginning of the Spanish-American War. Many of the hurricanes which formed in the Caribbean and Atlantic remained unreported until they moved in close to the coast of the United States weather net. An example of this occurred in September, 1874, only a few years after the formation of the weather service under the direction of the Chief Signal Officer of the U. S. Army. The storm of September 25-30 affected Florida, and, after being picked up on the weather chart, developed great force off the Carolina coast.

The first ship report of a hurricane was received by wireless on August 26, 1909, when the S.S. *Cartago* was off the coast of Yucatan. This part of the early system of ship reports was one of the twice-daily messages sent by wireless from all ships so equipped. Provision was soon made for reporting at intervening hours, and during periods of unusual weather. Thus, from the time of the first weather message from a ship at sea in 1909 until the present, the procedure of using ship observations has continued to improve. Ships at sea now take balloon soundings to obtain the direction and velocity of the winds aloft, and the use of the radiometeorograph on shipboard is a common form of weather observation.

Between the time of the invention of the weather map and the use of the wireless for reporting weather from ships at sea, much effort was put forward by research meteorologists to determine the exact nature of the hurricane. Reports from the laymen of the Islands were of little value. These reports described in detail the damage done, but during the storm only the approximate wind velocity could be indicated, and after the storm passed every effort went towards the repair of the damage. Some mention in the early reports brought out the fact that many hurricanes were associated with the advancement of definite cloud formations. The winds and temperatures received comment from some of the observers, for it was often calm and the weather was oppressive with the heat.

Roberts, writing in his introduction, gives an excellent lay description of the coming hurricane. This was the only information available to the population prior to 1900, and is the principle rule still used by many. "The signs of what is about to happen are unmistakable. Small cirrus clouds appear, drifting high and fast. They are followed by nimbus clouds, dark and ragged. The tides rise, because far away the advancing storm is whipping the water before it. The first phase may take an hour or two. The world has grown preternaturally still—and then the wind and the rain come roaring."

Father Viñes, of Havana, was one of the early students of tropical meteorology who made a concentrated effort to set up a series of forecast rules for the benefit of the public. He found that often when the hurricane is still at a distance the barometer sometimes rises above its normal heights. Soon thereafter the barometer begins to fall. The amount of fall at any given place depends upon the intensity of the hurricane and the direction of its movement with regard to the points of observation. The success of Father Viñes' system is noted in a letter describing the hurricane of 1928, nearly fifty years after the father went to work on the idea. Part of the letter by Charles R. Hartzell of San Juan follows: "The first indication of the storm was on Tuesday night of the 11th, when my barometer began to feel unusually healthy and went to what was plainly the anticyclone of a disturbance."

With reference to the barometer as an instrument to foretell the coming of the hurricane, it has also been used to determine the intensity of the storm—the common rule being that the lower the barometer, the greater the storm. From actual observations this has not always proven to be the case, but the record of some of the world's lowest pressures still gives rise to an exciting study. Our own country holds the record for the lowest reading of a shore-based barometer, but lower readings have been reported by ships at sea. Some of the low readings often referred to are listed here below:

	Inches
H.C. Ship *Duke of York*. Kedegree, 1833	26.30
Havana, Cuba. October 10 to 11, 1846	27.06
Ship *Favorita*, in harbor, Apia, Samoa, April 6, 1850	27.05
False Point, India. September 22, 1885	27.08
S.S. *Arethusa*, 13°55′N, 134°30′E. December 16, 1900	26.16
Basilan, Frank Helm Bay. September 25, 1905	26.85
S.S. *Sapoeroea*, 460 miles east of Luzon. August 18, 1927	26.18
Lower Matecumbe Key, Fla. September 2, 1935	26.35

Works by Gray, Dunn, Reid, and others have shown that the pressure is not a definite criterion of the storm intensity. The Miami storm of September, 1926, is generally conceded to be as severe as any storm ever to hit the United States. Studies of this storm are important because most of the record was made by trained Weather Bureau personnel using modern instruments. The barometer at the Weather Bureau office fell to 27.61 inches when the eye of the storm passed over the city. Another severe Florida hurricane passed over West Palm Beach in 1928, when a low reading of 27.43 inches of mercury was recorded.

Charles L. Mitchell, of the U. S. Weather Bureau, carried out an extensive research program to try to establish the average track and origin of the hurricanes which move into the Caribbean and the United States. He replotted the tracks of all storms between 1887 and 1922, inclusive. The daily weather maps of the Weather Bureau were checked to verify details, and ships' logs for the period were studied. Results of his study are still considered to be the best information available on the storm frequency, average tracks, and place of origin during each part of the season. Mr. Mitchell's findings can be summarized:

(1) The advance and decline of the number of storms during the season is definite. A hurricane season can be narrowed down to the three months of August, September, and October.

(2) Majority of the hurricanes developed

over the western Caribbean or far to the east of the Lesser Antilles.

(3) The influence of large high pressure areas on the movement of the hurricane is very marked.

(4) A hurricane will recurve to the north and northeast at the first opportunity regardless of the time of the year.

(5) When the hurricane describes a loop it is always to the left.

In recent history the New England hurricane of September, 1938, is still cited in many illustrations used in literature describing tropical storms. This storm, like many of its forerunners, had certain characteristics which make for interesting study. Its speed of movement from Hatteras to Long Island is a case of an unusual condition attending the storm. Other illustrations point up the fact that another storm, but of weaker nature, occurred in 1944; these two storms of 1938 and 1944 are set up for comparison with the two occurring in 1815 and 1821. Any study of hurricanes must then be narrowed somewhat if any approach is to make for intelligent evaluation.

It is not enough to describe the intensity of hurricanes in terms of loss of life, maximum winds, highest tides, lowest pressures, and many other features given to the hurricane. An accurate study of the four hundred year history is hampered because of the changing values of the units often set up for comparison. In most histories the attention is drawn to the loss of life, which can often be attributed to the high winds and tides accompanying the storms. One of the greatest catastrophes of all times is that referred to as the Hooghly disaster of 1737, when a furious cyclone destroyed 20,000 craft of all descriptions, and 300,000 people perished. In our own country the memorable Galveston Hurricane of 1900 is often recalled when a heavy loss of life is spoken of, for 6,000 persons gave up their lives in this storm.

To the mariner, layman, and others the history of the hurricane is only a backdrop to the present-day developments which aid him in protecting life and property from the fury of the hurricane. For this reason the last twenty years are the most important in the hurricane history. Some authors prefer to narrow the time interval down to the last ten years, arguing that it was not until 1935 that scientific study of the upper air was started in the Caribbean area. We might even go farther and say that it was the wartime effort of aircraft weather reconnaissance and microseismic work that began to pay the largest dividends in better hurricane protection.

Regardless of the time interval used in the history of the hurricane, it is necessary that we realize the stage of the present development to know how far the study has come in the last four hundred years. We have progressed from a point where there was no knowledge of the physical nature of the hurricane until the present day when it is possible to track them hourly across the ocean by use of aircraft, electronics, and microseismic devices. From the records of such methods it is possible to determine the size, shape, cross-section, and intensity of the storms. Weathermen are working on the assumption that forewarned is forearmed, and if the population can be warned of the coming storm the public can devise means for its own protection. This system of warning has been extended to all ships at sea, aircraft which may enter the storm area, and others who may see fit to use the information.

A short review of the two major wartime developments in the study of tropical storms is in order to complete the history of the hurricanes. Prior to the war, and extending from about 1935, considerable work had been carried out in the Caribbean area to extend the number of upper air observations. Pilot balloon soundings had increased in number so that a daily chart of the winds at all levels could be worked out to aid the forecaster in predicting the course of any hurricane. Radiometeorograph stations were expanded to give a better cross-section of the atmosphere in the vertical. The study of the upper air over the Caribbean is of major historical importance to the United States and to ships at sea which may be under the hurricane influence. During the hurricane season of 1944 work was started with aircraft observations in and near the hurricanes as they progressed along their track—the idea being that if an airplane could go out and spot the storm center, report the attending winds, and return to base, a running track of the storm could be plotted from each such observation. A flight could normally take off early in the morning from the nearest base to the hurricane, fly through the storm for observation purposes, and proceed to another field for safety. Another flight could carry out a similar procedure during the afternoon to give a minimum of two flights each day. Both the Army and Navy made experimental flights in 1944 which proved this to be a successful way of tracking the hurricanes. These flights were made with a good background of experience gained since the war began.

In 1945 the first tropical weather reconnaissance by the use of aircraft was established on a scheduled basis. For the Atlantic service twelve planes were made available—six from the Army Air Forces, and six Navy patrol bombers. These flights were co-ordinated in their observation work by the joint staff of Army, Navy, and U. S. Weather Bureau officials. Similar work was also carried out in the Pacific for the benefit of task force units and fleet commanders. . . .

Parallel to the aircraft reconnaissance program was the microseismic program whereby hurricanes would be tracked by the use of microseismographs. Microseisms related to hurricanes are thought to originate at the center of the storm over the sea. Father Macelwane, of St. Louis University, conceived the idea of a tripartite station consisting of three instruments in a triangle to take bearings on the hurricanes as they move over the ocean surface.

The Aerological Section of the Navy activated the Hurricane Microseismic Research Project late in 1943 [No longer in being.—Ed.] upon recommendation of the Joint Meteorological Committee. The first station was put in operation in 1944 at Guantanamo Bay, Cuba. The results of the Guantanamo station were verified by the weather flights in 1944 and 1945, making it possible to expand the work to cover the entire Caribbean, south Florida, and Corpus Christi. This work, too, has expanded into the Pacific as an aid in tracking tropical storms of that area.

To bring the history of the hurricanes up to date we have only to add the work of the hurricane reconnaissance squadrons and the microseismic project to the network of upper air observations. . . . With this close scrutiny of the area of storm formation it is unlikely that ever again will a hurricane breed, move through the ocean area, and affect the population of any Caribbean island, ship at sea, or United States city without adequate warning.

Appendix II : A Hurricane in Puerto Rico[1*]

By Lieutenant Commander O. H. Holtmann, U. S. Navy

Destructive hurricanes are not common in Puerto Rico. True, three have occurred in the past four years, two of these being classed as major disasters, the one of September 13, 1928, known as the San Felipe, after the saint's name of the day, and the one which occurred in 1932 on September 26, known as the San Ciprian. Prior to the San Felipe we must go back twenty-nine years to August 8, 1899, to find another record of a violent hurricane, the San Ciriaco. The great loss of life in that hurricane, three thousand, was due to the lack of facilities for prompt distribution of the warning to the rural population, and not to its particular violence, for winds of only 75 miles an hour were recorded in San Juan, which, however, was about 30 miles from the path of the center. The San Felipe in 1928 was the most destructive hurricane recorded in the West Indies. This storm formed near the Cape Verde Islands off the coast of Africa, and the center moved almost due westward until near Guadeloupe, French West Indies, then changed to west by north, and passed over that island where it caused great destruction. Passing a short distance south of St. Croix, where the lowest barometer reading was 27.50 inches, the wind velocity reached 90 miles per hour in St. Thomas, 50 miles north of the center. It entered Puerto Rico near Guayama (barometer 27.65) on the southeast coast and passed diagonally across the island in a west-northwest direction at an average velocity of translation of about 13 miles per hour. At San Juan an extreme wind velocity of 160 miles per hour was recorded when one cup of the anemometer blew away. The storm, however, increased in intensity for three hours after this. Previous to looking up the records I had asked Captain Evans, now in the SS. *Borinquen,* what the maximum wind velocity was in the San Felipe. He replied lightly, "Oh, about 200 miles an hour." The sailor's gift of exaggeration, I thought, and discounted this considerably. When one considers, however, that San Juan was about 30 miles north of the center, and that the lowest barometer reading in San Juan was 28.81, compared with 27.65 at Guayama, Captain Evans does not seem far from the truth. In Puerto Rico 300 were killed, and the property damage was estimated at fifty million dollars. This was the same storm which, continuing on to Florida, caused 1,836 deaths there.

Of the nine most destructive hurricanes which have occurred in Puerto Rico during the last hundred years, only one, that in 1867, occurred after the first of October. As September, 1932, drew toward its close, the people here, especially the fruit growers, who had thousands of boxes of

[1] The author wishes to make acknowledgment to Mr. F. E. Hartwell of the U. S. Weather Bureau, who furnished much of the information contained in this paper.

* From United States Naval Institute *Proceedings*—February 1933.

fruit on the trees, and needed one hurricane-free year to recover financially, began to lose their anxious expressions. Their hope was disturbed when at 9:00 A.M., on September 26, the Weather Bureau at San Juan issued the following bulletin:

> Tropical disturbance central this morning between St. Kitts and St. Martin. Should reach the Virgin Islands this afternoon and eastern Puerto Rico by midnight.

St. Kitts reported a barometric pressure of 29.68 and wind from southwest force 7. This last report was not in itself alarming, for while it definitely established the fact that a cyclone had formed, and that Puerto Rico would probably lie in its path, it was at that time of only moderate intensity. At Dominica, about 140 miles south-southeast of St. Kitts, the prevailing trade winds were not affected, that place reporting a normal barometer with wind from east, force 2, indicating a very small center. From the estimated center at 9:00 A.M. to San Juan was about 190 miles, and in that distance it could be expected that the intensity would increase greatly. By 9:00 P.M., St. Thomas and St. Croix reported 60 miles an hour gales and the Weather Bureau estimated that the center was somewhere between these two islands. Up to this time there was no indication in San Juan of anything unusual in the weather. At noon there had been a thundershower with heavy rain which lasted about half an hour after which the sky cleared. All "indications of the approach of a tropical hurricane" as described in one of our textbooks were absent, although at dark the center was only about 75 miles away, and no one with whom I talked had any oppressive feeling or "felt it in their bones."

With no land areas lying in its path the intensity rapidly increased until the vortex entered the east coast of Puerto Rico, at 10:00 P.M. of the twenty-sixth probably directly over the harbor of Ensenada Honda, where the steamer *Jean* of the Bull Line and the lighthouse tender *Acacia* had anchored. The *Jean* had both anchors down with 90 fathoms of chain on each in good holding ground, and her engines were turning over before the vortex struck. She dragged almost a mile in a southeasterly direction and went aground before the wind shifted. The bottom was soft and luckily she remained on an even keel. By lightering her cargo she was refloated without any serious damage the next day. The captain reported that the barometer was fascinating to watch, the needle vibrating over the extreme bottom of the scale (27.70 inches). The *Acacia* dragged all the way to the entrance of the harbor and when the wind shifted with the passing of the center, she dragged all the way back again, with her engines going at full speed, and finally went aground in the northern part of the harbor. The barometer on the *Acacia* reached a low of 28.00 inches.

In San Juan at 10:00 P.M., the barometer read 29.73, wind from the north 27 miles an hour. From then on the barometer dropped rapidly, and the wind rose correspondingly fast. At midnight the barometer had dropped to 29.50, wind northeast 62 miles an hour. Shortly thereafter the steel framework at the Weather Bureau, on which the anemometer was mounted, collapsed, but the meteorologist, Mr. F. E. Hartwell, estimated that at 12:50 the wind was blowing not less than 120 miles an hour, the barometer at that time reading 28.95. There was no lull for the exact center passed to the southward, while the wind shifted through east to southeast. Thereafter the wind dropped rapidly and by 4:00 A.M. it was all over, the wind shifting back to the east (normal trade wind direction) with a velocity of only 35 miles an hour, barometer 29.60.

At the height of the storm, the noise could best be described by the sound made by a locomotive starting a heavy train just before the exhaust merges into a continuous roar. The volume of the sound can be inferred from the fact that minor sounds, such as that made by a large limb striking across a wooden shutter, a garage door weighing perhaps 600 pounds being thrown against the back door, or a palm tree 3 feet in diameter at the base being blown onto the roof of the garage, were drowned out. . . . Sand and water were forced into seemingly impossible places. The "water-tight" globes around the light fixtures on our front porch, protected by a 2-foot overhang of the porch roof, were full of water although the roof, made of cemented tile, held and did not leak. Water was forced in under doors, through cracks, and around the window casings at such a rate that vigorous work with mops and towels was given up as useless.

Morning showed the extent of the disaster. Men had been at work since daylight clearing the roads of fallen trees, telephone poles, and all sorts of wreckage, and by 11:00 A.M. a car could get through most of the main highways. The island which had been so beautiful before was now in indescribable confusion. There was hardly a leaf left on the trees and shrubs still standing. Winter seemed to have come suddenly during the night. On one vacant lot in the Condado section, I had previously counted eighty-two coconut palms, all of them estimated to be over forty years old. After the hurricane I counted forty-two still standing. Over the countryside familiar landmarks were missing, their place being taken by others whose existence was not guessed before, since they had been hidden by shrubbery. In some sections there was not a building left wholly standing within 10 miles. Casualties were 225 dead, about 3,000 injured. Property damage will reach nearly thirty million dollars. The greatest loss of life was caused by the collapse of buildings which were supposed to be "hurricane-proof." Only the heaviest concrete or masonry buildings with tile roofs escaped undamaged. A peculiar effect of damage caused by air pressure as distinct from that caused by wind velocity was noted in the case of our garage which was of the same heavy construction as the house and, except for the doors, was almost air tight. The doors were torn off, one being thrown against our back door, moving a little north of east, while the other became jammed against a palm tree which had been blown down, and moved about northeast. The only winds of hurricane force were from north and northeast! Pressure inside had blown the doors directly into the wind.

The wind seemed to lose some of its violence as the vortex moved overland, being greatest where it entered the island on the east coast, and decreasing somewhat at the point where it left, on the northwest coast, near Arecibo, although this latter town reported the greatest proportion of deaths. Its total was swelled, however, by twenty-four men who took refuge in a new public garage, all of whom perished.

After leaving Puerto Rico the only report of winds of hurricane force was that from San Pedro de Macoris on the southern coast of Santo Domingo (90 miles per hour). No further report of the storm was received for some days and it finally entered British Honduras south of Belize.

Damage to shipping except small craft was slight. Beside the *Jean* and the *Acacia* previously mentioned, one ship in San Juan Harbor had her bridge and boats blown away, the three-masted schooner *Gaviota* was wrecked in the harbor and several pier buildings were badly damaged.

After a four-hour superficial survey of the affected area on the day after the storm, my reaction was "The San Felipe couldn't have been as bad as this, because it will take more than four years to remove the scars of this one." A month after the hurricane I was not so sure. In the tropics nature heals her wounds quickly. . . . Two weeks later there were new buds and shoots on the trees, the hibiscus was in bloom, and it was spring again.

Appendix III : Pitching on a Prayer*

Captain I. F. Wood, *Alcoa Runner*
Alcoa Steamship Company, Inc., Mobile, Alabama

This article . . . has been reprinted with the kind permission of Captain Wood. It first appeared in a 1956 issue of Polaris, *the quarterly magazine of the U. S. Merchant Marine Academy at Kings Point, Long Island.*

. . . As Master of a freighter I matched wits (unwittingly) with one of those devastating tropical cyclones. It was hurricane Carol of 1954, the one that toppled the steeple on Old North Church in Boston, snuffed out over 68 lives, and caused estimated damage of almost one-half billion dollars.

The ship, my command, is a C-2 type freighter built in 1944 at Wilmington, North Carolina . . . a single screw, turbine vessel, 459 ft. long, 63 ft. wide, and 37 ft. deep, having a raked stem and an elliptical stern. . . . Her normal speed is 16 kt. . . . She can carry a "payload" cargo of approximately 10,500 tons, which is an enormous amount of cargo for any type ship.

Ten thousand tons of bulk phosphate were loaded at Tampa, Florida on August 26, 1954, destined for one of the fertilizer plants at Norfolk, Virginia. When the ship departed from Tampa, she was loaded to her maximum draft of 28 ft., 6⅞ in.

. . . The pilot left at the Tampa seabuoy on August 26 at noon. A course was set south to pass between Rebecca Shoals and Dry Tortugas, the southwest extremity of the Florida Keys. The sea was smooth and the atmosphere clear.

That evening, the Radio Officer received the first U. S. Weather Bureau storm advisory on hurricane Carol. It read, "A small tropical storm has formed in the Atlantic and at 4:30 p.m. it was near 29.2°N., 76.5°W. or about 275 mi. east of Daytona Beach, Florida and moving northwest about 12 m.p.h. Strongest winds estimated 65 to 70 m.p.h. over a small area near the center. It is increasing slowly in intensity and may soon reach hurricane force. Northerly movement indicated for the next 12 to 24 hr."

. . . It was felt that the storm was so far in advance of the ship that there was only a remote possibility of overtaking it.

The ship passed between Rebecca Shoals and Tortugas at midnight, August 26 and all day August 27 the Keys and lower coast of Florida lay on our port side about 8 mi. off. The mighty Gulf Stream was flowing north with a velocity or strength of 3½ to 4 kt. The ship must have been in the axis of the Gulf Stream. When she was abeam of Miami she was averaging 19½ kt. The sky was blue with a few fleecy clouds, the sea was smooth, the temperature was 85 degrees, and the barometer was normal. There were great expectations for a record passage from Tampa to Norfolk due to the fine weather and the rampaging Gulf Stream.

At noon on August 27 the Radio Officer received Bulletin No. 4 on the hurricane:

* Extracted from *Mariners Weather Log* (U. S. Department of Commerce)—November 1963.

"Carol has remained almost stationary during the last 6 hr. but is probably moving very slowly 3 to 5 m.p.h. in a northerly direction. At 11 a.m. it was centered near 30°N., 76°W. or 325 mi. east of Jacksonville. Strongest winds estimated by aircraft this morning 70 to 80 m.p.h. extending 40 mi. from center. Gale force winds extend outward 50 to 60 mi. This small hurricane will probably show very slow increase in intensity and size during next 24 hr. Movement will continue in a northerly direction. Shipping should exercise extreme caution from present position northward to the Carolina coast next 36 to 48 hr."

Jupiter Lighthouse, just north of Palm Beach, was abeam at 4:30 p.m. August 27 and speed was reduced to 45 r.p.m. or "Half Ahead." I informed the responsible officers about the reduction in speed and told them that the purpose was to allow Carol to pass on to the northward of us.

From star observations taken early on the morning of August 28 it was found that in spite of the reduced speed, the vessel had traversed considerable distance during the night (with the aid of the Gulf Stream). A Weather Bureau advisory obtained at 5 a.m. located the hurricane about 300 mi. east of Jacksonville with highest winds estimated at 115 m.p.h. near the center and gales extending out 100 mi. The storm was expected to increase slowly in size and take a northerly path. At 6 a.m. the Chief Mate was called out and given orders to prepare the ship for heavy weather by putting storm plugs in all ventilators, extra lashing on the lifeboats, and securing plough steel wires across all hatches to the cargo holds.

At about 6 p.m. on the evening of August 28 the Weather Bureau report still described Carol as practically stationary. At this point the ship was in such a position that an important decision had to be made. She was between the hurricane and the mainland. The Gulf Stream was adding 2 m.p.h. to the engine speed. If the vessel was turned around and headed back for Miami, the hurricane might dally around further and a loss of three or four precious days on an already tight schedule might ensue. If she were pressed on to her destination, she might get caught in the hurricane—a deep loaded ship worth 5 to 6 million dollars, $135,000 worth of cargo, and 46 lives. . . . The ship was experiencing good weather and the barometer was normal, showing no evidence of the storm. A decision was made to proceed. If the estimated position of the hurricane were correct and it stayed stationary for the next 24 hr., the ship could be past Cape Hatteras and almost to Norfolk. . . .

Justification for the decision to proceed and apprehension for the safety of the command kept me on the bridge all that night. At midnight the barometer had dropped slowly from 30.00 to 29.80 in. but there was no alarm over that. By 5 a.m., however, the sea was increasing and the wind and seas were shifting toward the starboard beam. These C-2 ships roll very heavily when the wind and sea are on the beam. In a rough or heavy beam sea they roll severely enough to endanger equipment and fittings. At 6 a.m. the course was gradually changed from 40 to 70 degrees to ease the rolling.

The seas were growing longer and heavier and by 9 a.m. on August 29 they had become so violent that there was no longer any delusion about escaping the hurricane. The ship had been maneuvered into the path of the storm or dangerously near it. Plans were made accordingly to handle the ship with as much care and skill as could be mustered. The Chief Engineer was ordered to start pumping a slow trickle of fuel oil over the side by means of the bilge pumps. . . . There was a 50 gallon drum of so-called "Storm Oil" in the bows with outlets through the stem of the ship but the efficacy of that would have been somewhat equal to the effect of a bucket of water tossed at a house on fire. A steady flow of black fuel oil oozed out of the starboard side of the ship's hull all the time she was in the hurricane. No one will ever know how much that expended fuel oil helped to break the force of the seas encountered during the period of the hurricane.

. . . By "The Law of Storms" the rules and directions for one caught in my position were to put the wind on the starboard quarter and run for it. There are always exceptions to the rule and this was one of them. Had the wind been put on the starboard quarter, the ship would have been heading for the coast between Charleston and Cape Hatteras. . . . However, in addition to the impossibility of doing "what the book said," the deeply laden ship in my judgement would not have taken it safely before those solid walls of water. It was a great consolation at that period to hear a U. S. Coast Guard directive to Diamond Shoals Lightship to hold its position at all costs.

Throughout the storm the ship's bow was kept to the wind favoring the starboard side slightly because of the fuel oil discharge on that side. It was found experimentally that by varying the engine revolutions between 40 and 50 r.p.m., the vessel's bow could be kept into the seas preventing her from broaching broadside. The above mentioned revolutions were sufficient for the required purpose of steering directly at the seas yet seemed to give minimum aid to the destructive waves crashing over the decks.

Saturday August 29 was a wild, fierce day for all hands. The Log Book contained nothing but entries such as "mountainous sea," "heavy rain," "vessel rolling and pitching heavily," "wind force 11-12," etc. Ordinary routine broke down. Meals were only makeshift. Nobody slept. Nobody could. The rolling was too violent. It was a physical impossibility to keep oneself in a bunk. The wind blew so furiously that the spray and wind-driven rain actually beat the paint off the starboard side of the midship house down to the bare metal which is a thing rarely experienced in a lifetime at sea. Waves attained a height of over 40 ft. by the best approximation. Their size was so tremendous that it was frightening even to experienced seamen to look at them.

. . . We almost lost both lifeboats to the fury of the storm. Due to excessive rolling, the boats began to ride up in their davit tracks. All hands—engine, steward, and deck—turned-to to save the boats, and by much work, diligence, and improvisation the boats were kept safely secured in their cradles. It was grand optimism, our battle to save the lifeboats.

When the ship was taken off her normal course of 40 degrees she was put on a course of 70 degrees to ease the rolling. The course was gradually changed to the right as the wind and seas changed in that direction until eventually the ship was on a course of 200 or almost the reverse of her original course. About 30 minutes after midnight on August 29, I noticed that the wind and seas suddenly got into confused conflict. I was keeping a diligent watch on the steering to prevent the ship from getting broadside to those tremendous seas. It was deathly important that the helmsman not lose his vigilance or lapse into inattention. There was a perceptible movement of the wind and seas back to the left. The wind, rain, and spindrift were pelting the ship so unmercifully that it was impossible to see or look ahead except through the protection of the wheel house windows. I was fortunate to grasp the significance of that violent conflict of wind, seas, and rain and to quickly note the shift of the wind and seas back to the left. The ship was directed gradually back around an arc of the compass from 200 through north until eventually she was on a course of 270 or due west.

There was a time during the storm when

I doubted the survival of the ship if an abatement of the wind and seas didn't favor us. Those mountainous seas would surge at the bow and it seemed inevitable that they were going to sweep the decks clean and knock the midship house over the stern, but miraculously the ship would raise over the worst of the impact, shaking tons of spray back over the bridge, enveloping it in green water and blinding all vision for seconds.

. . . The ship's radar was kept in continuous operation during the hurricane and the whistle was dutifully blown at 2 minute intervals as required by law in low visibility. However, the rain and spindrift came in solid sheets at times and the whistle and radar were just something to keep us company. Disaster would have been inevitable had an unfortunate ship been on a collision course with us.

The weather began to abate rapidly after noon Monday, August 31. Eventually the ship was on a course of 270, due west toward the Carolina coast. At 6 p.m. on that day when about 25 mi. off the coast heavy fuel oil was pumped over the side to break the force of the tremendous swells and the ship was turned around to a southerly course and away from Carol and the shoals. An inspection of the decks was made for heavy weather damage. No major or heavy damage was found but considerable damage to small deck fittings and electrical cables was discovered. I lay down that night at 10 p.m. after 66 continuous hours of vigilance on the bridge.

Hourly readings of the barometer were recorded in the Log Book from the time the ship was abeam of Jupiter Light. A summary of the readings is:

August 27	6 p.m.	30.06 in.
August 28	Noon	30.00
August 28	4 p.m.	29.91
August 29	4 p.m.	28.74
(Lowest reading during storm)		
August 30	8 a.m.	28.81
August 30	Noon	29.24
August 30	4 p.m.	29.60
August 30	6 p.m.	29.78
August 31	8 a.m.	30.02

. . . Captain E. W. Malanot of MSTS recently wrote a very fine article called "Typhoon Doctrine" (see page 179) in which he advocates stopping the ship and letting her ride it out—engines stopped-dead ship. People have asked me why I didn't try the Malanot doctrine. I fear that Capt. Malanot's article will be misleading to some people, especially shoreside workers and inexperienced seamen. He mentions his draft of 12 ft. 2 in. forward and 17 ft. 3 in. aft. I can imagine that such a draft would be favorable in certain types of ships for laying dead in the sea before a storm. But the Malanot plan would be unsuitable for a heavy, deep-loaded ship like mine when loaded to her marks 28 ft. 6⅞ in.—one foot deeper than her builder's maximum draft due to that doubler plate welded to her sheer strake. . . .

Appendix IV : Riding Out a Philippine Typhoon*

By Commander R. P. Eyman, U. S. Coast and Geodetic Survey

"It's an ill wind that blows nobody good" is a trite old proverb that has been bandied about through the years and one wonders whether there is any basis of truth in it, especially at such times when you happen to be the "nobody" on whom the elements are venting their full fury. However that may be, I am sure that I can speak for the entire ship's company that thoughts, if any, along that line those of us on the *Pathfinder* may have had while riding out a severe typhoon in the northern Philippines, were most certainly centered on nothing but the forward half of that familiar quotation. In retrospect, of course, having safely weathered the particular storm, the full meaning of the saying began to dawn as the real value of such an experience was recognized—not, mind you, that there has ever been any desire to seek a more liberal "education" by repeating such an experience unnecessarily.

The U. S. Coast and Geodetic Survey ship *Pathfinder*, a flush-deck, coal-burning, steel steamer of 875 tons, had in her time weathered more or less successfully many typhoons. Built in Elizabeth, New Jersey, in 1899, she was sent from duty in Alaska to the Philippines in 1901 where she continued in operation until destroyed in early 1942 as a result of Japanese hostilities. Her triple-expansion engines of 846 horsepower gave her a speed of about 12 knots originally, but this had decreased through the years to a maximum of about 10 knots. The complement consisted of 9 officers (Americans) and a crew of 71 Filipinos.

On being ordered to command of the *Pathfinder* in the spring of 1932 instructions were received to resume hydrographic surveying operations, initiated several years previously, along the northwestern and northern coasts of Luzon. Work in these areas could, in general, only be carried on from March through October; that is to say, during the period of the tail end of the northeast and the major portion of the southwest monsoons—in short, the typhoon season. Having had, as a junior officer, several rather close brushes with typhoons during a previous three-year hitch in the Philippines and an experience or two with hurricanes while in command of a small vessel in Florida waters, the writer had developed a healthy respect for such storms and realized that no "gravy-train" assignment lay ahead of us for the next few months. Although the general condition of the 33-year-old *Pathfinder* was considered fairly good, there was no intention that we take too much for granted or deliberately pit her seagoing qualities against undue weather hazards, if such could be avoided gracefully. Suitable all-weather harbors were few and far between, being limited to that of Port Bolinao, 130 miles to the southward, and Port San Vicente, 100 miles to the eastward of Cape Bojeador on the northwestern tip of Luzon, with the additional possibility of

* From United States Naval Institute *Proceedings*—March 1945.

finding fair anchorage in the mouth of the Cagayan River at Aparri, approximately midway along the north coast.

By the end of June work off the northwest coast had been completed. This part of the project had been carried on steadily and without unusual incident. Weather had been generally good and the few times it became necessary, adequate shelter from the northeast could always be obtained by anchoring close under the lee of the shore. In fact, this part of the season might readily have been termed the "milk run." After making a few minor repairs, including the renewal of a number of defective boiler tubes and the installation of two tobin-bronze metal rods as diagonal bracing along the after bulkhead of the wooden pilothouse to stiffen it and remedy a slight "working" which occurred with the roll of the ship, we sailed from Manila on July 11 to take up our assigned work along the north coast of Luzon. By now the Northeast had subsided and typical Southwest conditions had set in with light variable breezes, smooth seas, and good visibility, the latter often exceptionally good just before a storm. There was a rather disturbing element present, however, that caused us to keep a very close watch on the weather. A trough of low pressure extending from the China Sea through the Balintang Channel and far out toward the Ladrone Islands had developed shortly before our arrival and showed no signs of filling up or moving on. As a matter of fact, it persisted for about a month. This trough or path of least resistance offered an enticing invitation to any disturbances that might be forming in the Pacific typhoon breeding grounds to set a course in our direction and hook up full speed ahead.

Except for a survey of the bar and mouth of the Cagayan River, made soon after our arrival, to determine whether access could be had to an anchorage within the river mouth—shoaling was found to preclude this—plans had long since been made to take the best possible advantage of every break in the weather to operate farthest from shelter and to reserve the near-by areas for operations during doubtful or suspicious periods when it might become necessary to seek refuge on rather short notice. In general, local weather conditions continued favorable and work was carried on with but one interruption until July 25, despite the fact the "lows," disturbances, and typhoons were being reported almost daily in practically all surrounding areas. On the 18th a typhoon had appeared in the Pacific about 300 miles to the eastward of Luzon and by the 20th was crossing that island only about 60 miles to the southward of our anchorage at San Vicente. It did not, however, affect us to any great extent aside from passing squalls with 50- or 60-mile breezes which, coming off the land, did not kick up any seas.

By July 25 a low that had shown up several days previously far out toward the Ladrones had finally graduated and was reported as a typhoon of light intensity some 300 miles to the eastward of central Luzon moving WNW—in our direction. Work was continued throughout the day, but indications of approaching trouble began to be manifest. Although the barometer had been below normal for some time, its diurnal oscillation now disappeared and the trend was definitely downward. The surface wind was a light NE breeze, but the high clouds indicated greater velocities. Clouds became thicker and squalls became more frequent and violent, and the breeze steadily freshened to moderate NE. Anchorage was made that evening in San Vicente outer harbor and a close watch kept on developments. During the night the barometer continued to fall gradually, squalls became somewhat more frequent but only slightly more intense, and the breeze continued to freshen from the NE. Weather reports received at this time did not indicate anything more serious than a low, or very light typhoon, still located about 200 miles out in the Pacific. By early morning of the 26th it was apparent that a typhoon of considerably more severity than previously reported was closing in on us and that it was now time to make all preparations accordingly. The ship was taken into the inner harbor and anchored in the center of the basin with the port anchor to which "false" palms had previously been attached. All awnings, dodgers, and loose gear were removed from above deck; small boats and other deck gear were secured with additional lashings; all rigging and stays were examined and set up taut; hatches not immediately necessary and skylights were covered and battened down; the second set of false palms was attached to the starboard anchor; the spare bower anchor was rigged over the side in readiness for dropping and the end of a coil of 12-inch Manila hawser broken out from the hold and made fast to it for emergency use—needless to add that a full head of steam was maintained and the engines kept warmed up for instant use.

Having completed all the above preparations by 0900, our situation was carefully reviewed. The ship was riding comfortably to one anchor in the center of a landlocked harbor, with swinging room of about 450 yards' diameter; the depth of water within this area was about 4 fathoms with mud bottom (which we hoped would prove of good holding quality— how good, we found out to our entire satisfaction later). Fringing the anchorage basin for an average width of about 400 yards were shoals and coral reefs, permitting a maximum scope of anchor chain of about 75 fathoms providing the mooring held in the center of the basin. The ground tackle consisted of two stock 1500-lb. bower anchors to which the false palms above referred to had been attached. These palms, four in number, were of 1-inch boiler plate about 2 by 3 feet, roughly triangular in shape, curved slightly to fit the regular palms snugly, and held in place by means of four 1-inch steel bolts; this extra equipment added about 250 lb. to each of the 1500-lb. bower anchors. The anchor chains consisted of 94 fathoms of 1⅛ inch on the starboard, and 120 fathoms of 1¼ inch on the port. The spare anchor was a stock type of 2079 lb. Since the *Pathfinder* had tendency to yaw badly when riding to a long mooring, it was decided to limit the scope first to 45 fathoms and later, if necessary, to pay out not more than 60 fathoms. We would ride to one anchor as long as possible, or until the course of travel of the storm could be more definitely determined, and then, when necessary, the second anchor would be dropped where it would serve the best purpose. The wind was still holding NE with no sign of shifting, indicating that we were on or very near the storm path, and latest reports placed the center only some 100 miles off. By noon, with the wind still holding the same direction and having steadily increased to about 50 miles per hour, squalls becoming more frequent and violent, and the barometer dropping steadily about .04 inch per hour, it was decided to delay no longer in dropping the second anchor and so, with the next lull, the starboard anchor was let go not too wide off the port anchor and both chains paid out to 45 fathoms and secured.

Anticipating a wakeful and rather busy night and since there seemed nothing else to be done for the time being, the writer thought a little siesta might come in handy, and so, with a competent watch set, time was taken out after lunch for a rest of about a couple of hours. The storm gradually but steadily grew more violent up to 1600, with practically continuous rain and a NE wind up to about 70 m.p.h. Shortly after 1600 the storm began to break in real earnest with squalls of increasing fury; seas began to pick up in our sheltered anchorage; darkness settled down and visibility became extremely limited. An officer was stationed in the eyes to keep the pilothouse advised of the lead of the moorings; the engines were used frequently to keep the ship's head into the wind and to ease somewhat the strain on the anchor chains, taking due care not to override them; the drift lead of about 15 lb. was soon replaced by another of about 25 lb., but this proved of academic interest only as the force of the wind on the bight of the line was so great that it was impossible to determine the relationship between the drift lead and the vessel; a Quartermaster was detailed to take occasional soundings off the stern to try to detect any incipient dragging. The fury of the storm increased in violence each minute. The gale howled and screamed through the rigging. The torrential rain was carried in horizontal sheets by the force of the wind which continued to rise in violent gusts—upon its reaching 100 m.p.h. the vanes of the anemometer were carried away, thus preventing further recordings. It was impossible to see beyond the bow of the ship. The waves in the harbor were whipped up to a height of 6 to 8 feet and the crests were blown off in a continuous foam of white spray. The pilothouse shuddered with each heavy gust and it was feared that it might be carried away. To face the wind and rain directly was practically impossible. Watching the lead of the moorings was an extremely difficult task. To make one's way about the deck it was necessary to cling to a rail or life line and, wherever possible, crouch below the bulwarks. A cautious turn about deck was taken at intervals to examine the security of gear and several times it was found necessary to break out the crew to strengthen some of the lashings—one of the launches was found partially lifted from its chocks. As the storm mounted in fury the barometer continued its nose dive; with each tapping of the glass the indicator hand dropped a hundredth or two with no sign of leveling off; each time the pilothouse door was opened the barometer needle would drop another point and a small hatch about 20 inches square in the center of the deck leading to a small storage compartment beneath was blown off.

Shortly before 1800 the force of the gale appeared to abate slightly and within a very few minutes rapidly decreased in violence to a moderate NE breeze. This sudden change in the situation called for immediate action as it was now evident that the center or eye of the storm was upon us and that the gale would break anew at any minute, probably from a new quarter. Anchor chains were hove in short in readiness to swing to the new direction when this could be determined, and a close watch was kept for the first signs of the rear half of the storm. The wind dropped to a very light breeze, veering slowly from ENE to NE to N × E and then died out completely. The rain ceased; the heavy clouds began to grow thin and finally broke away entirely revealing a beautiful blue, unclouded sky overhead with the last waning rays of the setting sun pouring through. This "fool's paradise" lasted for only about 20 minutes. The barometer had finally stopped its dive at 28.12″. The lull gave us an opportunity to make a quick survey of our situation which appeared to be so far so good; no serious damage was apparent and the anchors had dragged very little, if any.

Soon, however, thin cloud streamers began racing across the clear blue overhead from an ESE direction—this was the clue we had been awaiting. The ship was quickly swung around to this new heading and maneuvered to a position between the anchors with the engines. Within no time the surface breeze began to pick up from the ESE; the spare anchor was let go and the ship kicked astern to a mooring of 45 fathoms on the chains and with about the same on the Manila hawser—all were equalized as much as possible and secured. It was fortunate that this scope had previously been decided upon as it was now discovered that there was a badly frayed strand in the hawser only a few fathoms inboard when all had been secured. These arrangements had barely been completed before the storm again began to gather real headway—dense clouds had blotted out the blue sky, heavy rain squalls had returned, and the breeze freshened in quickly succeeding gusts. Within less than 10 minutes the gale had mounted to its previous intensity and finally exceeded its former fury. This half of the storm was somewhat more nerve-racking than the first half, what with complete pitch-black darkness except for lightning flashes and with the greater severity and frequency of the gusts which reached an estimated velocity of well over 100 m.p.h.—probably 115 to 125. It was much more difficult to keep a close watch on the lead of the moorings, but the officers assigned to this duty did a magnificent job and as result of their directions, signaled by flashlight, the ship was kept headed into the gale with fairly uniform tension maintained on the moorings. It was now simply a matter of riding it out and hoping that the anchors would hold and nothing give way, as it seemed that having passed through the forward half of the storm in a matter of a couple of hours, the rear half, although much more intense, should not last too long. The barometer by this time was rising rapidly.

Shortly after 2200 a slight diminution in the strength and frequency of the gusts was detected. By 2230 the decrease was decidedly noticeable and from then on the force of the gale rapidly declined until shortly before midnight when only a very gentle breeze was blowing from the SE. Although the sky remained partly overcast, the rain squalls became infrequent and of no particular violence. We had apparently weathered the storm safely and, although it was too dark to determine whether we had dragged anchors, at least we were afloat and not piled up in the midst of mangrove bushes.

The morning of the 27th broke cloudy, with light southerly to southwesterly breezes and intermittent light rains. It was noted that the ship had dragged not more than 15 or 20 yards during the entire storm, and the answer was soon disclosed. What was thought to be a mere routine job, that of picking up the anchors, was commenced about 0700, but was not completed until about 1130; the windlass lacked power to break them out and it was necessary to heave each one short in turn and steam ahead full speed to accomplish it. When brought to the surface the reason was at once apparent; the anchors had dug well into the bottom and each brought up a load of hard-packed mud weighing probably a half ton or more and requiring the use of shovels and pike poles along with the hosing to remove it and clean the chains and anchors. There remained no further doubt about the holding qualities of this particular anchorage.

A careful examination revealed that except for the pilothouse we had suffered practically no damage. The force of the wind had opened many seams in the tongue-and-grooved framing of the forward part of the pilothouse in a number of places by as much as one-half inch and it is firmly believed that the diagonal bracing, referred to before, was all that saved it from being carried away. Some of the awning stanchions were somewhat awry and a few spreaders were missing, but otherwise everything was intact. The deck was littered with all sort of debris from ashore—small branches, palm fronds, sand, pebbles, etc. During the night we had also captured a large sea bird of the goony family which had been blown in from sea by the gale and had sought refuge on the deck of the ship. On making a turn of inspection one of the officers had suddenly and literally come face to face with this tall, black, strange creature in the dark behind one of the hatches and with the help of several crew members had succeeded in placing it in a large crate; it was liberated the following morning when it appeared to have recovered from its exhaustion.

Although the barometer had leveled off, it still remained below normal and the weather continued unsettled and did not clear as quickly as might be expected once the typhoon had passed. Local conditions, however, were not unfavorable and work was resumed on the 28th and carried on until the afternoon of the 29th, when, fuel and supplies running low, it was necessary to return to port. On rounding Cape Bojeador southbound, somewhat to our surprise we ran into a strong SW gale with rough seas. Despite reports indicating much better weather ahead, conditions became worse the farther south we proceeded and the barometer crept lower. Not caring to subject the gallant old vessel to unnecessary punishment and strain, we eased ahead through heavy head seas and pulled into San Fernando about noon on the 30th and awaited there more favorable weather. On the morning of August 2 our voyage was resumed and arrival at Manila was made on the morning of August 3, completing not-soon-to-be-forgotten experiences with tropical disturbances that had ranged through the full scale within a period of less than one month.

It was with genuine regret that some months later orders of detachment were received and my relief came aboard to take over command, but my tour of duty in the Philippines was finished and it was now time to return to the States and leave the vessel in the hands of my successors, who invariably held the same high regard and feeling of warm attachment for the faithful old ship. The *Pathfinder* continued on surveying duty in various areas of Philippine waters until the advent of the present war. Having recently returned to Manila, she was slightly damaged in one of the first Japanese bombing raids, but was able to proceed to Corregidor with valuable records. Off Corregidor she suffered further damage from bombings during the last days of December, 1941, and had to be beached. The last report, transmitted in March, 1942, indicated that the 43-year-old veteran was among those listed for total demolition to prevent her falling into enemy hands. Although the long career of this vessel had thus come to an inglorious end, her name is perpetuated in a new ship completed that same year. The new *Pathfinder* is a modern all-steel vessel of 1,900 tons, especially built and equipped for surveying work in the most remote regions, and, since her completion, has been steadily engaged on just such duties, frequently in enemy-held territories, the accomplishment of which has had a vital part in the successful operations of our naval and military forces in carrying the war ever closer to total victory.

Appendix V : The Typhoon Lady*

By Lieutenant (j.g.) Robert J. Lauer, U. S. Navy

Quite a few naval vessels weathered quite a few typhoons during the recent war in the Pacific, but I feel that very few could approach the record of survivals per typhoons encountered established by the U.S.S. *San Jacinto*.

The little CVL met her first tropical storm west of the Palau group of islands in September of 1944. However, her baptism was mild, and she only skirted the storm, suffering nothing worse than two days of heavy rolling and pitching.

The disdain for typhoons thus engendered in the minds of her ship's company was shattered abruptly in December of that year when, as a unit of Task Force 38 supporting Army operations in the Central Philippines, she became entangled in the worst marine disaster the Navy had known to that time. That infamous typhoon which capsized two destroyers and several smaller craft leaving no survivors, came upon Admiral Halsey's carrier task force at precisely the time when it could not retire to safer waters. The Army had just landed troops on Mindoro Island and needed every bit of air support the Navy could provide. The decision to remain in the area and assist the invasion until the last moment was, no doubt, a difficult one to make but the success of the operation surely justifies the losses suffered at the hands of the elements.

Having been warned beforehand of the approaching storm the *San Jac* had set her typhoon bill. All moveable gear had been lashed down. Additional lines from aircraft on the flight and hangar decks were made fast to the securing grommets in the deck. Elevators were lowered to increase the GM and the "Little Queen" waited for the blast.

The intensity of the storm could never have been imagined beforehand. Winds of over 100 knots and seas 70 to 80 feet high lashed at the ships of the formation. To ease the ferocious pounding the course was adjusted to place the ships in the trough of the seas, and speed was reduced to the minimum required to maintain steerageway. All hands stood by their respective spaces to insure their security.

It was a terrifying sight to watch the gigantic breakers on the crest of the seas looming up, sometimes as much as thirty degrees above the horizontal, as the ship slowly rolled through forty degrees or more. It would have been suicide to venture onto the flight deck. It speaks well for the flight deck crews that every plane located there held fast.

However, down on the hangar deck it was another story. The TBMs, the heavier planes, had been struck below. The hangar deck could just hold the ten of them in two rows of fives. They were lashed down with all the manila available, since this took place before the advent of steel cable securing lines. At the peak of the storm the weight of the planes pulling against the

* From United States Naval Institute *Proceedings*—June 1949.

lines under the excessive roll proved too much to hold. A plane broke loose and was hurled into the one beside it, breaking it loose from its moorings. With every roll of the ship the two would careen across the deck and smash against the bulkheads, carrying away everything in their paths—spare engines, belly tanks, tractors and other articles of gear.

An heroic effort to harness these murderous missiles and secure them to the bulkheads was led by the air group commander, Commander Gordon Schecter (who was subsequently lost in action over Kyushu), but it soon proved impossible and exceedingly hazardous.

In a very short time other planes had been cut loose by the wreckage of the first two, and within a half hour the entire hangar deck was a mass of flying carnage threatening sure death to anyone who dared enter into it. A spare engine near the port side aft broke its securing lines and, on the starboard roll, flew across the hangar deck and passed through a steel roller curtain as though it were a piece of paper.

Gasoline and oil covered the deck. A spark was struck, and flames leapt up in the middle of the hangar. The hangar deck sprinkling was turned on, and hoses and foam were played on the fire from the openings to the cat walks. This, combined with the torrential rain and sea water from openings punctured in the sides soon produced a virtual flooding of the hangar deck. Then machine gun ammunition from the planes began to go off, fired by the flames and by being struck with the flying debris.

The four uptakes and most of the ventilation piping led through the hangar. These were ruptured and carried away by the hurtling wreckage, and the salt water flowed into the openings, flooding many compartments below and freezing or shorting out much electrical equipment as well as cracking one boiler drum.

Spare steel plates, carried in the elevator pits, slithered back and forth mercilessly gouging the elevator pistons and rendering them inoperative. All ventilation and most lighting was lost. It seemed almost inescapable that if the storm continued the ship herself must soon be abandoned. Then, like a last minute reprieve, the storm began to abate. The fires were extinguished and flooded compartments pumped out.

As the ships re-formed toward sunset of the third day of the storm the *San Jac* seemed but a ghost of her former self. She had power for propulsion and steering, but not much else. The humid air, and the lack of ventilation and light, coupled with the sickening motion as the ship rode over the long swells characteristic of a passed typhoon, made living below unbearable. Meals in the wardroom, with tables and chairs lashed to the deck or bulkhead and with smashed china on the deck, consisted of unsatisfactory rations of cold sandwiches and rancid coffee. Two feet of water stood in the crew's mess and had to be pumped out before meals could be served. Everyone aboard was exhausted.

But then as if by magic morale skyrocketed. The rumor swept the ship in a few minutes. Naturally she would return to the States for repairs! To a crew which has been in the forward area continuously for eight months the news was like manna from Heaven.

The task force retired to Ulithi, where representatives of ComSeron 10 inspected the damage. The scuttlebutt continued. The crew nervously awaited the word that the next port would be San Francisco. But the fates decided otherwise, and a sister ship, the *Monterey,* also badly damaged, was elected to return to the States after being cannibalized by the *San Jacinto.*

The "Little Queen" went alongside a repair ship where she stayed for ten days; the ten days over Christmas. The crew of the repair ship, the *Hector,* worked untiringly during that time, twenty-four hours a day, and did a miraculous job of overhaul. By the first of the year the ship was as good as new and at sea, about to participate in the next operation.

Needless to say the information that the ship would remain in the forward area did very little to help morale, particularly in view of the fact that every man aboard had been quite certain he would soon be home, and because the repairs were affected over Christmas week. It would be difficult to depict the depth of the gloom which settled over the ship Christmas morning when, instead of reveille, *Deck the Halls with Wreaths of Holly* was piped over the P.A. system.

However, it was but a matter of a few days at sea until all hands felt normal again. A week after leaving Ulithi the task force was in the South China Sea where, it was hoped, the remnants of the Jap fleet could be ferreted out and destroyed. Three weeks later the task force steamed through the Balintang Passage, leaving the South China Sea without having seen a major Jap naval vessel, a smooth sea, or the sun.

Operations during the month of February placed the *San Jacinto* successively off Iwo Jima, Tokyo, Iwo again and Tokyo again, and finally in the Inland Sea, headed for strikes against the Nagoya area. These latter strikes were cancelled, however, for icy winds from the north were churning up such violent seas that aircraft operations were deemed too hazardous to attempt. The task force then retired to Ulithi to prepare for the nightmarish three months of strikes against the Japanese mainland and the Nansei Shoto, including the invasion and securing of Okinawa. The only pleasant feature of that whole affair was, ironically, the weather.

The respite from the fury of the elements terminated in June, however, when the task force, operating southeast of Okinawa, encountered, simultaneously, two typhoons which culminated their northeasterly movement by combining at approximately the same geographical location occupied by Task Force 38.

The task group of which the *San Jacinto* was a unit bore the brunt of the storm and experienced a rare accomplishment in passing through the eye of the westernmost typhoon. At that point the wind and sea moderated considerably, and a patch of blue sky could be seen overhead. The barometer had reached a record low of, I believe, about twenty-seven inches. Of course as the cyclone moved on, its force again built up to a maximum, and then gradually subsided. The violent effects of the storm were felt for approximately forty-eight hours.

During the greater part of the storm the ships were headed so that the wind was on the port bow and the sea on the starboard bow. Turns were made for five knots, but actually the ships barely moved through the water.

The *Pittsburgh,* alongside the *San Jacinto,* was so violently shaken that she lost her bow. The *Duluth* narrowly escaped a like fate, but suffered such severe pounding that she was forced to retire to Guam to have her bow strength members repaired. The *Bennington* and *Hornet,* the CVs of the group, suffered identical casualties when both of their flight decks forward collapsed over their forecastles after the supporting members had been carried away by the seas. It was interesting a few days later to watch them launch aircraft off the after end of their flight decks. Practically all scouting planes topside on the battleships and cruisers were either carried away or hopelessly battered.

The *San Jacinto* rode out this typhoon with a minimum of damage, having suffered nothing worse than several buckled deck plates and cracked seams, and a number of slightly bent deck stringers. However, the personnel experiences were quite harrowing. For a period of about five hours as the ship rode out the mountainous seas the effect was as though some titanic crane were lifting the bow higher and higher, until it seemed the ship would surely break her back, and would then release it and permit it to drop with a frightening crash and shudder.

After the storm had passed the *San Jac* found herself, with the exception of the 20 mm sponsons forward, in full operating condition. She had learned from experience how to conquer a typhoon.

As the Japanese envoys were signing the terms of surrender aboard the *Missouri* on September 2, 1945, the U.S.S. *San Jacinto* was headed for San Francisco, after seventeen months of uninterrupted operations against the enemy, for which she later received the Presidential Unit Citation.

It seemed that even in peacetime the little CVL could not escape the violence of the sea. In December of 1945, while engaged in the Magic Carpet operations, she was en route from San Francisco to Manila, P.I. Her first ten days out were devoted to a continuous fight against violent seas in the form of breakers which hammered night and day against her starboard bow and succeeded in tearing away most of her stack supports and smashing number one stack into uselessness.

Permission was obtained for the return trip to avoid the rough great circle route and to take the longer, balmy route due east from San Bernardino Straits to just south of the island of Hawaii. By that time the *San Jac* had seen her last rough seas, although she narrowly missed the fatal tidal wave which two days after her arrival in San Francisco devastated the city of Hilo on Hawaii.

For two years now the *San Jacinto* has been resting peacefully, inactivated, alongside a pier at the Alameda Naval Air Station, and the worst weather she experiences is the fog rolling in through the Golden Gate and condensing on her flight deck. She seems to have settled into a sort of blissful retirement.

Appendix VI : Typhoon Doctrine*

By Elmer W. Malanot

It is a fact that even the most experienced seaman in command of a ship sooner or later finds himself facing some problem where he will have to consult a textbook on seamanship; and in most cases, *Knight's* or *Crenshaw* will have the answer. This is not the case, however, if a man is looking for advice on what to do, and how to do it, when a typhoon or a hurricane has caught up with him.

Cyclonic storms are fully described, and also the method on how to avoid them, and how *sailing ships* can avoid them, but I have never yet seen anyone suggesting what to do, and how to do it, should one find himself caught in such a storm.

It is my assumption that the men who wrote these books on seamanship did not have sufficient experience passing through tropical storms and that their source of information was not complete enough to enable them to write on this subject.

If a man who has had such experience were to suggest stopping the engines when the wind in a cyclonic storm reaches a force of sixty knots and doing nothing until the center has passed and the force of the wind has reduced to about sixty knots, people would be skeptical. It would be argued that in stopping the engine, the ship would fall off into the trough and capsize; therefore to avoid falling off into the trough considerable power ought to be used to keep steerage way. It is not realized that this power hurls his ship against mountains of waves, like ramming one's head against a stone wall, and that in this way he would damage and perhaps even destroy her unwittingly.

Now, where is the trough in a cyclone that captains try to avoid? To find an answer to this question, we should study the relation between wind and waves, and between waves and ship. Sea waves are formed by prolonged action of the wind. Once a sea wave is formed it will continue for some time, and travel great distances even when the wind that caused it has calmed down or changed its direction. Sea waves that travel at a speed of ten to twenty knots do not carry the water with them, as it is only the form that advances while the particles of water have a different motion from that of the waves.

This phenomenon can be seen in tidal estuaries, with waves rolling in from the sea against an ebb tide and floating objects being carried out to sea by the tide against the inward passage of the waves. This can also be seen in a ship dead in the water. The onrushing crest of a wave will not hit the ship; it will only lift it. But if the ship is moving, it will be the ship that hits the wave. This point of *who is hitting whom* is most important and is seldom understood by the average seaman.

In a typhoon, all observers concur in their reports, near the center the sea is confused. This confusion is caused by the changing direction of the force that forms

* From United States Naval Institute *Proceedings*—July 1955.

the waves. In cyclonic storms, a wind blowing from north, for instance, will build up a heavy sea from that direction. Then, after the wind backs to northwest, and so on, there will be several distinct series of waves within the same area at the same time. This constant change of wind direction is what makes the terrible confused seas which just cannot be avoided no matter what the ship does.

Hurricane in the Caribbean

Many years ago when I was a comparatively young shipmaster, in command of a freighter of about 16,000 tons displacement on her maiden voyage in the Caribbean Sea, we had a hurricane warning. It was my first experience with a tropical storm, and with the help of *Bowditch* I put the ship on a course which would have brought it into the safe semi-circle out of the hurricane's path.

The motorship being deeply laden to the Plimsoll mark, worked heavily. Towering seas crashed on deck, and there was some damage to life boats and deck fixtures. During the voyage we had some difficulties with our diesel auxiliaries, and now the engineer reported that he had two out of commission, and it did not take long until the last auxiliary also failed. This of course stopped the main engine, and we were left at the mercy of that hurricane roaring down on us with winds of 120 knots. Visualizing what might happen to us when our ship fell off into the trough of those gigantic waves gave us quite a scare.

When we gradually recovered from that first shock and the ship was lying dead in the water, it seemed strange that the action of sea and wind decreased considerably and that there was no more smashing of waves on deck, although the center passed directly over us, with a short dead calm and blue sky. Then again came the tearing and screaming wind from the opposite direction. It seemed to us miraculous that there was no damage to the ship at all after the engine broke down. It took us some time to patch up the auxiliaries, and then we proceeded; but this lesson I have remembered all my life.

Later I read an account in the *Reader's Digest* about such a miracle happening to a destroyer during Admiral Halsey's battle with a typhoon in December, 1944. That destroyer had its boiler rooms swamped in the early stages of the storm, and although lying helpless and dead in the water, it survived the typhoon without further damage, while three other destroyers who had their 60,000 horsepower working to the last capsized with the loss of every man on board. Those poor men died without knowing that it was suicide to use their powerful engines against the infinitely greater power of nature.

After V-J day I was assigned to the shuttle service between Manila, Okinawa, and Yokohama, then later extended to include Guam and Korea. From January, 1946, until May, 1954, except for five months' leave, I was in the typhoon area constantly and logged an average of two major typhoons a year. This offered excellent opportunity to study and observe typhoons in all their phases.

Typhoon *Libby*

In October, 1948, we were caught in Naha, Okinawa by *Libby*. We could not leave the port because there was a fifty-knot wind blowing three days before the typhoon struck the island, permitting no traffic to pass through the narrow channel of that harbor. There were, besides my ship (the USAT *Pvt. George J. Peters*), two Liberty-type ships manned by Japanese. These two ships tore from their moorings and piled up on Engineer Island in Naha harbor, heavily damaged. My ship had only superficial damage to the railings caused by one of the Liberties, for our moorings held. *Libby* was unusual because the center stopped over the island for about five hours, a most unusual occurrence. On October 3, 1948, the wind (northerly) increased gradually during the night to 120 knots, then in the morning of the 4th it suddenly dropped to a dead calm. The barometer was at about 28.40 and the sun was shining. At 1500 hours it started to blow from the SSW, then west and full force of about 140 knots from the WNW. At midnight the barometer started rising and the wind decreased to about forty knots at daylight, October 5. Those days in Naha, Okinawa, were dangerous, and I promised myself never to be caught again in a port by a typhoon.

Typhoon *Gloria*

The following year on July 22, 1949, the USAT *Pvt. George J. Peters* was again in Naha when *Gloria* came over the Island. This time we put out to sea and drifted with stopped engines while the center passed us at a distance of forty miles. The wind was up to 120 knots, the waves gigantic, of a height estimated to about 50 feet, but the ship floated like a duck, and there was not the slightest damage or even discomfort on board.

Typhoon *Ruth*

The most violent and destructive storm I have experienced in those Far Eastern waters was typhoon *Ruth*. At that time I was in command of the USNS *Sgt. Joseph E. Muller*, a C1-MAV-1 class vessel of 4,500 tons displacement. When we arrived in Okinawa, *Ruth* was 500 miles south, heading toward Formosa. We entered Naha harbor at 0700 hours on October 12, 1951, discharged our passengers, and were starting cargo operations when at 1000 hours the evacuation of Naha harbor was started. We proceeded to sea at about 1400 hours; our draft was 12′-2″ forward and 17′3″ aft. Our metacentric height (GM) about 4.84 feet.

Captain Henry T. Jarrell, USN, our passenger from Yokohama to Formosa, was offered quarters ashore by the Senior Naval Officer on Okinawa, but he returned to the ship to observe our typhoon-fighting tactics of which he had heard reports from officers ashore.

From Okinawa we proceeded to the island of Kume Shima, approximately fifty miles west from Naha, to take shelter behind the island until we had a clear indication of the proximity of the typhoon. We stayed near this island from 1840 hours on October 12 to 1747 hours on October 13. At that time we abandoned our shelter and let the easterly wind carry us out to sea with engines stopped. The barometer was falling rapidly, indicating the approach of the typhoon's center. The wind's force was about fifty knots, increasing continuously. There was a heavy swell from the south into which the ship was heading, being broadside to the easterly wind. The vessel was naturally rolling and pitching, but without taking any sea or spray on deck. Gradually the wind shifted counterclockwise, the barometer fell steadily, and the wind increased rapidly. By 0300 hours October 14, when we were at the closest point to the center; the wind had shifted to north; the force of it estimated at about 140 knots; the sea from all directions of a height estimated at 50 feet, and the barometer reading 28.02, the ship heading 090 degrees, drifting in a southerly direction with engines still stopped. At 0400 hours, the barometer began to rise, and the wind shifted gradually to NW and continued to back through W to SW. At 1030 hours the wind decreased to fifty knots, and we proceeded at full speed toward Naha, where we arrived at 1600 hours.

From various indications, it was established that the center of the typhoon had passed between Naha and our position, approximately 25 miles to the east from our vessel. From the damage caused by this storm on land and to shipping, *Ruth* was considered one of the fiercest typhoons Okinawa has had. Although the center passed us at a distance of only about 25 miles, we never experienced the slightest concern for the safety of our ship; we had no sea breaking on our decks, no jars or shocks to our hull, and no damage whatsoever. The only thing we did was to stop the engine and offer no resistance and no fight to the elements. Two fair-sized Japanese freighters which also evacuated Naha took shelter and anchored between the islands of Kerama Retto. During the night of October 13-14 they dragged their anchors and were smashed to pulp on the rocky shores of the island. Each was a total loss.

The following conditions were typical while drifting in typhoon *Ruth:*

a. Visibility zero to 300 yards.
b. Drifting at speeds of a maximum of five knots and in a semicircle of which the diameter was thirty miles.
c. Pitching heavily and rolling about 35° with a few rolls up to 43°. In general the vessel adapted itself easily to the heavy confused sea. It did not ship any water except for a very few times in the heaviest rolls when a few inches of water above the gunwale were scooped off the crest of a wave. This was like an overflowing motion with no force or violence attached to it, would have gone to a depth not over a man's ankle, and would have amounted to less than a few hundred gallons. It certainly would not have damaged or disturbed any sort of deck cargo secured in the normal manner.
d. Radar and Loran were operative and gave us our position when required.
e. The floodlights were kept alight on deck, and thus the behavior of the ship continuously observed.

. . . the heavy swell in the vicinity of the center of a typhoon is always from a direction different from that of the wind. In our case, the wind was from WNW, still about 100 knots, while the ship pitched heavily to a swell from the NNE.

This is the point that should be impressed on seamen who believe that if they stop the engines, their vessel will fall off into the trough of the sea with disastrous results. The fact is that there exists a trough from all sides, and the only way to minimize the danger of this force is by not opposing it. If inexperienced commanders of small vessels with high power do not fully realize this point, they will, in their anxiety to avoid this trough, hurl their vessels against the gigantic power of nature and succeed only in their own destruction.

I am fully convinced now that a seaworthy ship of any size or draft with an adequate metacentric height, will be safer with stopped engines than by using them. Any book on seamanship teaches us how to calculate the metacentric height, and how dangerous a sluggish roll or, to the contrary, a quick jerky roll can be; and it teaches us also how easily both can be corrected: the sluggish roll by adding some weight in the bottom tanks, and the other by removing some weight from the bottom tanks. In a typhoon there is sufficient time to correct the stability of the vessel and also to obtain a safe margin of seaway. And so once a seaman, after getting sea room, has learned not to use any power in a cyclonic storm, he will have done all that can be done for the safety of his vessel and his crew, and will further find that he can then take any typhoon in his stride.

I think that this theory of not opposing such a force and of stopping the engine is one of the greatest importance to combat vessels, and especially destroyers, which for tactical reasons cannot concentrate on avoiding a typhoon, but have to stay at their post until the last minute, as during operations east of Luzon on December 18, 1944. I am convinced that if those commanding officers had been instructed, on the approach of that typhoon when the wind had risen to more than sixty knots, to stop their engines and just drift, there would have been no casualties and no damage.

I believe that it is high time to break with the old tradition of fighting a typhoon and introduce a new concept of riding it out, with stopped engines, drifting more or less broadside to the wind, which in a typhoon is always from a direction different from that of the biggest and most destructive waves. It will take courage by the Commanding Officer of any vessel to order the engines stopped, when close to the center of a typhoon, and even more courage to order a ship to test this "passive resistance," but I am sure that if this method is adopted, it will save money, lives, and valuable property, especially in time of war.

It is suggested that in this respect, the following simple instructions should be put in textbooks and typhoon doctrines:

When the center of a tropical storm approaches you, get thirty miles or more sea room around you, assure yourself of the ship's stability and watertight integrity, and when the wind force rises to sixty knots or more and the sea becomes confused, stop your engines, drift and wait for the storm's dangerous area to pass before you proceed.

Appendix VII : Discussions, Comments, Notes

TYPHOON DOCTRINE*

In the July, 1955, *Proceedings* there appeared Elmer W. Malanot's "Typhoon Doctrine," an article which stimulated considerable comment from *Proceedings* readers. . . .

(Editor's Note: Mr. Malanot stated . . . that he had been inspired to write of his own experiences after having read an account of the typhoon which hit the U. S. Third Fleet off the Philippines in December, 1944. Soon after the Malanot article appeared, the Naval Institute learned of a Pacific Fleet Confidential Letter written by the then Commander in Chief, Pacific Fleet, Admiral C. W. Nimitz, and dealing in detail with the lessons of that disastrous typhoon. With the cooperation of Fleet Admiral Nimitz and the Office of the Secretary of the Navy, the Naval Institute here publishes Admiral Nimitz' letter, now declassified.)

18 February 1945

PACIFIC FLEET CONFIDENTIAL LETTER 14CL-45

From: Commander in Chief, U.S. Pacific Fleet.

To: PACIFIC FLEET and NAVAL SHORE ACTIVITIES, Pacific Ocean Areas.

Subject: Damage in Typhoon, Lessons of.

1. On 18 December 1944, vessels of the Pacific Fleet, operating in support of the invasion of the Philippines in an area about 300 miles east of Luzon, were caught near the center of a typhoon of extreme violence. Three destroyers, the HULL, MONAGHAN, and SPENCE capsized and went down with practically all hands; serious damage was sustained by the CL MIAMI, the CVLs MONTEREY, COWPENS, and SAN JACINTO, the CVEs CAPE ESPERANCE and ALTAMAHA, and the DDs AYLWIN, DEWEY, and HICKOX. Lesser damage was sustained by at least 19 other vessels, from CAs down to DEs. Fires occurred on three carriers when planes were smashed in their hangars; and some 146 planes on various ships were lost or damaged beyond economical repair by the fires, by being smashed up, or by being swept overboard. About 790 officers and men were lost or killed, and 80 were injured. Several surviving destroyers reported rolling 70° or more; and we can only surmise how close this was to capsizing completely for some of them. It was the greatest loss that we have taken in the Pacific without compensatory return since the First Battle of Savo.

2. In the light of hindsight it is easy to see how any of several measures might have prevented this catastrophe, but it was far less easy a problem at the time for the men who were out there under the heaviest of conflicting responsibilities. The important thing is for it never to happen again; and hence, while it is impracticable herein to go into all the factors involved and the experiences undergone, some of the outstanding lessons will be discussed.

3. Possibly, too much reliance was

* From United States Naval Institute *Proceedings*—January 1956.

placed on the analysis broadcast from the Fleet Weather Central, Pearl Harbor. Weather data was lacking from an area some 240 to 300 miles in diameter (where the storm was actually centered); and the immediate signs of it in the operating area were not heeded early enough. Groups of the Third Fleet tried to avoid the storm center, but neither radically enough nor to best advantage, since their information as to its location and path was meager. Fleet damage and losses were accentuated by the efforts of vessels and subordinate commanders to maintain fleet courses, speeds, and formations during the storm. Commanding officers failed to realize sufficiently in advance of the fact that it was necessary for them to give up the attempt, and give all their attention to saving their ships. There was a lack of appreciation by subordinate commanders and commanding officers that really dangerous weather conditions existed, until it was too late to make the preparations for security that might have been helpful.

4. The following conditions were typical during the typhoon:

(a) Visibility zero to a thousand yards.

(b) Ships not merely rolling, but heeled far over continually by the force of the wind, thus leaving them very little margin for further rolling to leeward.

(c) Water being taken in quantity through ventilators, blower intakes, and every topside opening.

(d) Switchboards and electrical machinery of all kinds shorted and drowned out, with fires from short circuits. Main distribution board in engine room shorted by steam moisture when all topside openings were closed to keep out water.

(e) Free water up to two or three feet over engines or fireroom floor plates, and in many other compartments. It apparently all came in from above: there is no evidence of ships' seams parting.

(f) Loss of steering control, failure of power and lighting, and stoppage of main propulsion plant. Loss of radar and of all ability to communicate.

(g) Planes on carriers going adrift, crashing into each other, and starting fires.

(h) Wind velocities and seas that carried away masts, stacks, boats, davits, and deck structures generally, and made it impossible for men to secure gear that had gone adrift, or to jettison or strike below topside weights when the necessity had become apparent. Men could not even stay up where they would have a chance of getting clear of the ship.

(i) Maneuvering up to the time of sinking, in the attempt to maintain station, by all ships that were lost. DEWEY, saved by apparently a narrow margin, had given up the attempt.

(j) The storm "taking charge" and making impossible various evasive and security measures which might have been effective at an earlier stage.

(k) Testimony that the ships lost took a long roll to leeward, varying from 50 to 80°, hung there a little while, and then went completely over, floating a short time before going down.

5. The following tabulation does not purport to be the whole story, either for the ships mentioned or for the Fleet as a whole. It does, however, show that some ships, although of the same class as those lost, and undergoing the same punishment from the weather, survived nevertheless. It also indicates some differences in their condition and in the measures taken. Nobody can say, however, how far the outcome was due to these conditions and measures (or lack of them) and how far to blind chance.

6.

Class	All of FARRAGUT Class				Both FLETCHER Class	
	Hull	Monaghan	Dewey	Aylwin	Spence	Hickox
Outcome	Sunk	Sunk	Survived	Survived	Sunk	Survived
Fuel on hand (app.)	70%	76%	?	80%	15%	14%
Water Ballast	No	No	Yes	?	Very little	Fully ballast'd
Fuel to high side	No	No	Yes	Yes	No	?
Cond "A" taken	Yes	?	Yes	?	No	?
Top weight jettisoned or below	No	?	Yes	?	?	?
Free water in ship	Yes	?	Some	Yes	Yes	Some
Rolled and recovered	70°	?	75°	70°	Hung at 50° then capsized	70°

7. Various weaknesses were brought to light in our forecasting and dissemination of weather information, in structural details which permitted flooding with consequent loss of power, short circuiting, etc., and in the stability of some of our destroyers. Measures to correct these faults are being taken as far as possible. Yet the Commander in Chief, Pacific Fleet wishes to emphasize that to insure safety at sea, the best that science can devise and that naval organization can provide must be regarded only as an aid, and never as a substitute for the good seamanship, self-reliance, and sense of ultimate responsibility which are the first requisites in a seaman and naval officer.

8. A hundred years ago, a ship's survival depended almost solely on the competence of her master and on his constant alertness to every hint of change in the weather. To be taken aback or caught with full sail on by even a passing squall might mean the loss of spars or canvas; and to come close to the center of a genuine hurricane or typhoon was synonymous with disaster. While to be taken by surprise was thus serious, the facilities for avoiding it were meager. Each master was dependent wholly on himself for detecting the first symptoms of bad weather, for predicting its seriousness and movement, and for taking the appropriate measures, to evade it if possible and to battle through it if it passed near to him. There was no radio by which weather data could be collected from over all the oceans and the resulting forecasts by expert aerologists broadcasted to him and to all afloat. There was no one to tell him that the time had now come to strike his light sails and spars, and snug her down under close reefs or storm trysails. His own barometer, the force and direction of the wind, and the appearance of sea and sky were all that he had for information. Ceaseless vigilance in watching and interpreting signs, plus a philosophy of taking no risk in which there was little to gain and much to be lost, was what enabled him to survive.

9. Seamen of the present day should be better at forecasting weather at sea, inde-

pendently of the radio, than were their predecessors. The general laws of storms and the weather expectancy for all months of the year in all parts of the world are now more thoroughly understood, more completely catalogued, and more readily available in various publications. An intensive study of typhoons and Western Pacific weather was made over a period of many years by Father Depperman at the Manila observatory, and his conclusions have been embodied in the material available to all aerologists. What *Knight* and *Bowditch* have to say on the subject is exactly as true during this war as it was in time of peace or before the days of radio. Familiarity with these authorities is something no captain or navigator can do without. The monthly pilot charts, issued to all ships, give excellent information as to the probable incidence and movements of typhoons. Stress on the foregoing is no belittlement of our aerological centers and weather broadcasts. But just as a navigator is held culpable if he neglects "Log, Lead, and Lookout" through blind faith in his radio fixes, so is the seaman culpable who regards personal weather estimates as obsolete and assumes that if no radio storm warning has been received, then all is well, and no local weather signs need cause him concern.

10. It is possible that too much reliance is being placed on outside sources for warnings of dangerous weather, and on the ability of our splendid ships to come through anything that wind and wave can do. If this be so, there is need for a revival of the age-old habits of self-reliance and caution in regard to the hazard from storms, and for officers in all echelons of command to take their personal responsibilities in this respect more seriously.

11. The most difficult part of the whole heavy-weather problem is of course the conflict between the military necessity for carrying out an operation as scheduled, and the possibility of damage or loss to our ships in doing so. For this no possible rule can be laid down. The decision must be a matter of "calculated risk" either way. It should be kept in mind, however, that a ship which founders or is badly damaged is a dead loss not only to the current operation but to future ones, that the weather which hinders us may be hindering the enemy equally, and that ships which, to prevent probable damage and possible loss, are allowed to drop behind, or to maneuver independently, may by that very measure be able to rejoin later and be of use in the operation.

12. The safety of a ship against perils from storm, as well as from those of navigation and maneuvering, is always the primary responsibility of her commanding officer; but this responsibility is also shared by his immediate superiors in operational command since by the very fact of such command the individual commanding officer is not free to do at any time what his own judgment might indicate. Obviously no rational captain will permit his ship to be lost fruitlessly through blind obedience to plan or order, since by no chance could that be the intention of his superior. But the degree of a ship's danger is progressive and at the same time indefinite. It is one thing for a commanding officer, acting independently in time of peace, to pick a course and speed which may save him a beating from the weather, and quite another for him, in time of war, to disregard his mission and his orders and leave his station and duty.

13. It is here that the responsibility rests on unit, group, and force commanders, and that their judgment and authority must be exercised. They are of course the ones best qualified to weigh the situation and the relative urgency of safety measures versus carrying on with the job in hand. They frequently guard circuits or possess weather codes not available to all ships; and it goes without saying that any storm warnings or important weather in-

formation which they are not sure everybody has received should be re-transmitted as far as practicable. More than this, they must be conscious of the relative inexperience in seamanship, and particularly hurricane seamanship, of many of their commanding officers, despite their superb fighting qualities. One division commander reports that his captains averaged eight years or less out of the Naval Academy, and this is probably typical.

14. It is most definitely part of the senior officer's responsibility to think in terms of the smallest ship and most inexperienced commanding officer under him. He cannot take them for granted, give them tasks and stations, and assume either that they will be able to keep up and come through any weather that his own big ship can; or that they will be wise enough to gauge the exact moment when their task must be abandoned in order for them to keep afloat. The order for ships to be handled and navigated wholly for their own preservation should be originated early enough by the seniors, and not be necessarily withheld until the juniors request it. The very gallantry and determination of our young commanding officers need to be taken into account here as a danger factor, since their urge to keep on, to keep up, to keep station, and to carry out their mission in the face of any difficulty, may deter them from doing what is actually wisest and most profitable in the long run.

15. Yet if the O.T.C. is to be held responsible for his smaller vessels, he must be kept aware of their conditions, and the onus of this rests on the commanding officers themselves. Each of them must not only do whatever he is free and able to do for his ship's safety, but must also keep his superiors in the chain of command fully informed as to his situation. If there is anything in his ship's particular condition or in the way she is taking the weather that worries him, he should not hesitate to pass the information to his seniors. To let this be regarded as a sign of faintheartedness is to invite disaster, and seniors should indoctrinate their commanding officers accordingly. Going still further, it has been shown that at sea the severity of the weather may develop to a point where, regardless of combat commitments of the high command, the situation will require independent action by a junior without reference to his senior. This becomes mandatory if grave doubts arise in the mind of the junior as to the safety of his vessel, the lives of its crew, and the loss of valuable government property and equipment.

16. The commanders of all echelons in the Pacific Fleet will impress upon their subordinates the necessity for giving full consideration to the adverse weather likely to be encountered in the Western Pacific, particularly the presence of tropical disturbances and the formation and movement of typhoons. In this connection, each commanding officer should refresh himself on *Knight* and *Bowditch*, not only as to the "Laws of Storms," but also as to ship-handling in heavy weather. In order to know what outside weather reports are broadcast and what he should be getting, each commanding officer should be familiar with *Radio Weather Aids to Navigation* (H. O. 206), and its confidential supplement, H. O. 206-C-S(A). This publication, as well as the *Navy Weather Forecast Code No. 1* (CSP-946) should be on all DD's and DE's, *etc.* Even more important, a commanding officer should check up on his own ship's system of handling dispatches to make sure that every incoming dispatch about prospective weather is viewed and understood by himself or some other officer with experience enough to grasp its significance. It should by no chance get buried in files and overlooked. This applies even more strongly to local observations. Preoccupation with the job in hand, or a desire not to disturb

the skipper, should never result in disregard of a rapidly falling barometer.

17. Steps must be taken to insure that commanding officers of all vessels, particularly destroyers and smaller craft, are fully aware of the stability characteristics of their ships; that adequate security measures regarding water-tight integrity are enforced; and that effect upon stability of free liquid surfaces is thoroughly understood. For preparing the ship against expected heavy weather, the basic written authorities are:

(a) *Damage Control Book* for ship concerned.
(b) Ballasting Instructions issued by the Type Maintenance Administration concerned.
(c) Notes on Stability of Ships in a Seaway (Pacific Fleet Maintenance Confidential Letter No. 7-44).
(d) *Booklet of Inclining Experiment Data* for either ship or class. Issued by BuShips.
(e) *Damage Control Instructions* (FTP 170-B).
(f) "Derangement of Electrical Equipment caused by Ventilation Conditions." *BuShips Bulletin of Information,* No. 12, p. 9.
(g) *Stability and Compartmentation of Ships* (C&R Bulletin No. 14).

18. In conclusion, both seniors and juniors must realize that in bad weather, as in most other situations, safety and fatal hazard are not separated by any sharp boundary line, but shade gradually from one into the other. There is no little red light which is going to flash on and inform commanding officers or higher commanders that from then on there is extreme danger from the weather, and that measures for ships' safety must now take precedence over further efforts to keep up with the formation or to execute the assigned task. This time will always be a matter of personal judgment. Naturally no commander is going to cut thin the margin between staying afloat and foundering, but he may nevertheless unwittingly pass the danger point even though no ship is yet in extremis. Ships that keep on going as long as the severity of wind and sea has not yet come close to capsizing them or breaking them in two, may nevertheless become helpless to avoid these catastrophes later if things get worse. By then they may be unable to steer any heading but in the trough of the sea, or may have their steering control, lighting, communications, and main propulsion disabled, or may be helpless to secure things on deck or to jettison topside weights. The time for taking all measures for a ship's safety is while still able to do so. Nothing is more dangerous than for a seaman to be grudging in taking precautions lest they turn out to have been unnecessary. Safety at sea for a thousand years has depended on exactly the opposite philosophy.

C. W. Nimitz

Commander C. R. Calhoun, USN—While I take no issue with the general theory that it frequently may be necessary in severe weather to give a ship its head (and in fact this is the gist of the now-famous Nimitz letter on the subject), I do wish to clarify Mr. Malanot's paragraph dealing with the Third Fleet's experience.

As Commanding Officer of the USS *Dewey* on 17-18 December 1944, I believe I can claim the distinction, along with my *Dewey* shipmates of having survived without capsizing, the greatest roll ever recorded by any U. S. naval vessel. I further believe that the *Dewey* probably came closer to disaster than any other ship which survived that storm, and feel that whoever wrote in *Reader's Digest* of a destroyer which "had its boiler rooms swamped in the early stages of the storm," was probably speaking of the *Dewey*. If this is the case then the use of the *Dewey* to illustrate Mr. Malanot's doctrine was an unfortunate choice. At no time during the entire storm were the *Dewey* boiler rooms "swamped" to the point where fires were extinguished, nor did she ever lose main propulsion power, nor did she lie "dead" in the water, for her port engine was kept one-third ahead throughout.

At about 0900 on the morning of 18 December 1944, I found it necessary to come left to avoid collision with the USS *Monterey*, whose Commanding Officer had just reported her dead in the water to fight hangar deck fires. The 40-degree course change thus made placed *Dewey* in the trough (there was no trouble in finding it—we were in it!) and also put the ship "in irons," with the wind generally on the port beam. Despite every possible engine and rudder combination, it was found impossible to move the ship's head more than ten degrees either way. I did discover that we rode just a bit more comfortably with the wind about one point abaft the beam and hence kept the port engine running at one-third ahead, with right 20 degrees rudder throughout the remainder of the storm. Although the seas became more confused as we approached the center, there still was a basic trough, and *Dewey* remained steadfastly wallowing in it.

To attribute *Dewey's* survival entirely to Fate, is to do an injustice to the gallant men whose efforts saved her. During countless giant rolls to starboard, the main forced draft blower intakes were submerged, and an estimated 500 to 1000 gallons of water poured into the firerooms. A brave little band of engineers, who never gave their personal safety a thought, remained in the firerooms, and by superb damage control techniques kept the flooding under control and maintained boiler steam pressure. I have the feeling that without the use of the port engine the ship would have been lost, for she came so close, so many times, to capsizing, that the loss of the slightest advantage would have been enough to tip the balance the wrong way. As I have already pointed out, the ship rode slightly better with steerageway and right rudder.

A few other facts might be of interest. The *Farragut*-class destroyers were known by their captains to possess a very small margin of stability. We were perhaps more stability-minded than any other skippers in the fleet, and we took unusual precautions to prepare our ships for heavy weather. The fact that two of them were lost was no surprise to those of us on the *Dewey*. Although the records are not now available to me, I believe it will be found that *Hull* (and possibly *Monaghan* also) did not have "60,000 horsepower working to the last," but rather capsized after her engine spaces had had to be abandoned! In my own personal opinion, the chief factor which contributed to the loss of the *Hull* and *Monaghan* was an inherently low margin of stability.

Contributing to (if not the direct cause of) the loss of the *Spence*, was her unballasted condition until in extremis when, if my memory serves me correctly, a last minute attempt was made to ballast her, and free-surface effect was thus added to the problem. *Spence*, a 2100 tonner, entered the typhoon with something like 10,000 gallons of black oil and no salt water ballast. Small wonder she was lost, and I doubt that the use of her engines had much to do with it!

As it stated in my second paragraph, I take no issue with the general "doctrine" proposed by Mr. Malanot. U. S. destroyermen have long recognized the advantage of letting the ship fight the storm as she seems to want to fight it. But let's correct the impression that *Hull, Monaghan,* and *Spence* were lost because their skippers pounded them against overpowering seas, or that *Dewey* survived simply because she had her boiler rooms swamped. It just didn't happen that way!

CAPTAIN JOHN V. NOEL, JR., U. S. NAVY.—All blue water Navy men will enjoy Mr. Malanot's *Typhoon Doctrine.* He has described sound procedures for handling a ship in very heavy weather, although the practice of drifting with engines stopped is not considered as unorthodox as Mr. Malanot would have us believe.

A ship anticipating having to weather a typhoon or hurricane should first look for a lee, an island, or peninsula that will provide shelter. The alternative, of course, is to seek maximum sea room. This latter practice was axiomatic in the days of sail because of a seaman's dread of being caught on a lee shore as the wind shifts. It is still a consideration, particularly for a low-powered vessel, but should serve mainly to emphasize the need for understanding the structure of a tropical cyclone so that you can anticipate and understand the shift in the direction of the wind. Many small ships have survived severe tropical cyclones by seeking the shelter of land, moving to new positions as the wind shifts. This shift in wind, together with the first signs of the barometer checking its plunge and starting to rise, is usually a most heartening sign and shows that the storm is moving away from you.

If there is no lee to be found the storm must be weathered in the open sea. There are no hard and fast rules to guide a commanding officer faced with a rising wind and a falling barometer. Certainly, up to a point, maximum speed should be made away from the estimated path of the storm. It is normally more comfortable to head downwind, and this course should be chosen when practicable. Eventually it may be found advisable to head up into the sea at minimum speed. The lowest speed at which steerageway can be maintained should be used. In a twin-screw ship the engines may have to be employed separately to keep the ship headed into the large storm waves. There may come a time, in very high seas, when the ship persists in falling off and nothing can be done to avoid coming around into the trough of the sea. Under these circumstances many ship captains have stopped their engines and allowed their ship to drift. In most ships the drag of the propellers will cause them to lie with the wind and sea on the quarter. Some ships are reported to drift remarkably comfortably in this position, with little solid water coming aboard in spite of heavy rolling and pitching. This is natural since there is little resistance on the part of the ship to the action of the waves. As Mr. Malanot so aptly expresses it: "for a ship dead in the water the onrushing crest of a wave will not hit the ship, it will only lift it."

REAR ADMIRAL S. C. RING, USN.—In substance, Mr. Malanot by virtue of the experience which he has gained, recommends lying to as the best maneuver to avoid damage on the approach of a tropical storm. He would resort to that action only under stated conditions and only after assuring himself of the stability of his vessel.

I do not gainsay the efficacy of the action proposed in handling *single* ships although personally I have never tried it. On the other hand, it seems pertinent to me to recognize that in many instances vessels confronting typhoons (or better yet, maneuvering to avoid them) are travelling in company and have a definite tactical mission to perform.

There can be little doubt but what the blind execution of an assigned mission is unwarranted in the face of forecasts which would produce weather damaging to the material integrity of the command. Adequate weight must be applied by the commander to probable gains to be acquired through the execution of his orders versus the danger of damage to his vessels incurred by immediate fulfillment of his assignment. Axiomatically, the commander should advise his immediate superior of his conclusions in the premises. A factor worthy of consideration is that weather is, by and large, impartial in its application and if weather conditions are such as to preclude successful execution of the mission assigned one force, it is highly probable that the anticipated maneuvering of the opposing force will be likewise affected.

Basically, the Task Force Commander should so maneuver his force that in the absence of compelling reasons to the contrary, he will avoid placing his force in weather-jeopardy. Such maneuvering may involve the acceptance of minor damage as the Force attempts to escape the path of the approaching typhoon. But, except as a last resort, the integrity of the command should not be lost. Obviously the commander should be guided by the manner in which all vessels of his Force are responding to the weather and he should so adjust the course and speed of the Force as to minimize the damage incurred, minimize the hazard to personnel, and maximize the chances of successful completion of his assigned mission.

Granted that each individual commanding officer bears the burden of responsibility for the ultimate safety of his vessel, and that under certain conditions he may elect to lie to in order to minimize the damage and danger to his ship and crew. But I believe that it is the responsibility of the Senior Officer Present so to maneuver his charge that the extreme conditions are not met. I further believe that danger is increased rather than lessened if the commander orders his group stopped. In a formation, operating to avoid typhoon damage, components of the formation should maintain steerageway in order to minimize danger of collision, and should remain as well prepared as possible to continue the execution of the mission assigned when weather permits.

Captain H. B. Lyon, USN.—In 1952, as Chief Inspector for MSTSNORPACSUBAREA, I had the opportunity to take passage from Sasebo to Pusan and return with Captain Malanot in the *Sgt. Joseph E. Muller*. Captain Malanot regaled me with this typhoon doctrine of his and his many, many experiences which bear it out.

He is a delightful person, with old-style continental manners, with a lifetime of sea experiences, and a burning desire that the discoveries that he has made may be shared by all ship masters everywhere. To talk to him is to be convinced.

At the time that I met this fine gentleman, he was at a loss to know how to get his findings the type of recognition that he thought they deserved. I have endeavored, from time to time, to promote Captain Malanot's principles in discussion and in local articles. However, I have discovered two things:

(1) A second-hand story carries little weight.

(2) A person with some typhoon experience is usually so completely overawed by the tremendous destructive power and majesty of a typhoon that it is difficult for him to think objectively about it.

Any Captain, in skirting a typhoon, will derive a great deal of security from the knowledge that "Typhoon Doctrine" imparts. It will give him assurance that he may survive even if his engines fail at the critical moment. However, Captain Malanot believes that his method is the only safe method of traversing a typhoon area, if that condition becomes a necessity.

Needless to say, I was very pleased to see his article in print. It is my sincere hope that more data by more people will become available to support his position. It must be available somewhere for the Gloucester Fishermen have been doing this for centuries.

Lt. Cdr. A. G. Graham, USNR (Ret.), Master, SS *African Enterprise*, Farrell Lines.—With regard to Mr. Malanot's "Typhoon Doctrine" I am qualifiedly in accord, having had one experience of breaking down in violent weather with the astonishing and comparatively favorable results which he describes.

The "doctrine" has one weakness and that is the heavy rolling that results. I know nothing about naval vessels, but a cargo ship loaded with bulk ore could shift her cargo and founder (as happened recently to the *Mormackite*) and a passenger vessel could seriously injure too many paying customers (most of whom seem to have a personal GM of 0.1 at sea) for the good of future business.

I have found that a modern merchant vessel, with its raked stem and flaring bows, can be hove to differently than was possible with the straight-stemmed and flat-bowed ships of pre-war days. For what it may be worth, I will quote from an article I wrote for our house organ, *Farrell Lines News* (January, 1955) on the subject of heaving to in heavy weather.

"When heaving to, head to sea, the recommendation usually found in seamanship books is to keep the sea from two to four points on the bow. The purpose of this is to give the bow a chance to rise to the oncoming sea instead of poking her nose right into it. I disagree with this recommendation, however. I have found it best to head directly into the sea. With a gale force wind on the weather bow, you have to keep way on her to prevent her from falling off into the trough, and the stronger the wind the more way you require. By heading directly into the weather you can cut speed to a bare minimum, which means fewer—and much less forceful—seas coming aboard. Another advantage is that it reduces rolling materially. I have followed this practice for years and am convinced of its superiority.

"In Hurricane Carol we made good 1½ miles in 5 hours (by accurate Loran fixes) which is as close to stopped—the ultimate ideal—as one could wish. I estimated the wind at 90 knots (it was clocked over Long Island a few hours later at 94 knots and attained 117 knots in gusts at Block Island according to the August, 1955, *Pilot Chart*) and the seas at 50 feet high. These seas were about 1000 feet, or roughly two ship's lengths, from crest to crest and were remarkably steep. This gave the ship no time to recover from the down slope of one sea before the next one was on her, and there were times when I thought she would stick her nose into it clear up to the bridge. Being all but stopped, however, gave her the few seconds needed to get her bow started upwards before it was engulfed, and we took only the tops of the breaking seas aboard. We suffered some damage forward, but it was of a minor nature.

"Prior to this I had been running before it at slow speed (though we still made 7 knots) simply because it was a 'fair' wind for us to New York. Finally, however, a sea pooped us heavily and carried away the hand steering wheel, the docking telegraph, and a few other useful items, and I figured it was time to turn around and face it like a man.

"Turning around under such conditions is always a heart-in-mouth proposition. This time I waited for 40 minutes (after alerting the Steward and Engine Departments to stand by for some heavy rolls) for a 'smooth' of sufficient extent to make the turn. Oddly enough, such smooths come along at intervals in the worst of blows. When it did come I gave her full ahead and hard left and crossed my fingers. She made the 180 degree turn in time to cut her to slow again before the first big one hit her. It snapped off our jackstaff and crushed a ventilator, but that was all, and my heart went back where it belonged. Running too long is an error and should be avoided if possible. However, lack of sea room or other considerations may make it necessary at times."

The above is applicable only when wind and sea are from the same direction, of course, but that is the normal situation. In or near the eye of a storm, however, with dangerous seas coming from different directions, it is impossible to avoid heavy rolling. Under such circumstances, I can appreciate the wisdom of stopping and drifting, as is recommended in "Typhoon Doctrine."

John F. Wilson, Master, SS *Pioneer Lake,* U. S. Lines.—In the past ten years I have commanded five C-2 vessels, in the North Atlantic, South Pacific, and U. S. East Coast to the Far East. I am presently master in a C-2 in the Far East service, having spent the past eight years in the service. My early training was as an officer in passenger vessels such as the *Leviathan* and *Washington*; there I observed how the outstanding masters of those vessels handled them in major storms. Those masters had been trained by other outstanding men. From this experience I have concluded that no rules can cover all vessels. All depends on the vessel's size, her cargo, how it is stowed, and her handling qualities which only her master can know.

In a loaded C-2 I fear rolling more than pitching or pounding. Experience has taught me just about how much working my vessel can stand. As long as possible, I "weather a gale," by maintaining steerageway on 40 RPM, with a man at the phone and a throttle watch, to increase should she go off. We make every effort to dodge the typhoons, but sometimes get caught—even with our booms up. I can control my vessel's behavior as far as pitching is concerned, but rolling, the cargo may shift or other circumstance beyond my control start forces to work that may cause my vessel to founder. If my C-2 were light, then I would act differently, fearing pitching and possible cracking.

Like Captain Malanot, my greatest experience was in the Caribbean Sea. It was September, 1933. I was an officer in the T.E.S. *Virginia* (32,500 tons displacement). We were caught in the center of a hurricane; the barometer, corrected, fell to 27.40; the wind force was more than 125 MPH, proved by the fact that the anemometer on Swan Island, near us, broke down at this force. The ship went out of control; the rain, driven by the wind, cleaned the paint off steel stanchions, as though they had been sandblasted. Rain water ran over the thresholds and under the outside doors, spilling into the main lobbies and cascading down the main stair cases. Our heads rang as though we had taken quinine tablets. To breath more easily, men in the fireroom had to remain near the floor plates. The hurricane bridge rails were folded back to the deck by the wind. Lifeboats were torn from their chocks and smashed against the boat deck house. I attempted to go down the outside ladder, and my 140 lbs. was virtually supported by the wind; I had to haul myself down. Fantastic and freakish damage was sustained. Since that time whenever I hear of surface winds of over 100 knots, I feel skeptical.

If a moderate size vessel is caught in a major storm there is not much choice but to stop, let her find herself, put out oil, if necessary, and hold on. In December, 1952, in the Pacific, I "guessed" correctly; a C-3 did not. We were near one another, running parallel to a storm center which weather reports from Guam stated held winds of 180 knots. The center was 400 miles south of us. *Pilot Chart* tracks showed most of these storms moved due west. Except one, that went right across our course. I slowed down, taking the chance that I might be held up for days. The C-3, larger and faster, continued on. The next day we exchanged weather reports. He was out of control, in high confused seas, winds of over 100 knots, and 150 miles from the center. He lost 165 tons of deck cargo over the side. As the Army tanks worked loose they moved his deck winches, this permitted water to enter number one hold and flood it. 250 miles east, I was having force 7 to 8 winds and heavy seas and swells, but no damage.

An authority has explained that swells are not raised by the hurricane winds revolving about the center of the typhoon, but rather by the pulsatory vertical motion of the lower part of the typhoon, causing "pumping" in the barographs and acting as a kind of piston on the water, thereby causing concentric waves to radiate outward from the center. And so, the swells in the vicinity of the center are always from a different direction than the wind.

If I were asked for advice by a young shipmaster I should say, "When possible—and I know it is not always possible—stay at least 250 miles from the center of a major storm, or at least clear of the confused seas, in order that you may escape."

But whatever decision you may make, if you get into trouble, you may be sure that someone, who was *not* there, will come up with something you *should* have done.

Robert E. Herbert, Jr.—The introduction and conclusion of Mr. Malanot to his "Typhoon Doctrine" in the July, 1955, *Proceedings* causes nearly as much of a confused sea as the storms he writes about.

My experience has shown the propeller to be a very good drogue, and under normal loading the ship's bow and forward areas present more windage than the after half. Therefore, the average vessel will lay to and ride comfortably with the wind and sea on the quarter. In steam-powered vessels, I have had the engines ticking over dead slow to keep them warmed up for emergency. I refer to the old "ups-and-downs" as I have not had any turbine ship experience. With diesels, of course, you just start them when you want them, though on one ship I was in, it was a gamble as to how much air she'd use before that old Atlas would fire.

It is a good article for the modern day, mechanically-powered mariner and gives further weight to my constant argument that the graduates of our maritime schools, including the Naval Academy, would be better officers and ship handlers if they had some training in sail such as the Coast Guard has. In their case, they have made a good beginning, but I agree with Allan Villiers that the *Eagle* is not used nearly enough.

UP TO DATE TYPHOON DOCTRINE*

(Editor's Note: The writer of the following comment is the author of the article "Typhoon Doctrine" which appeared in the July, 1955 *Proceedings* and brought forth numerous comments which were published in the January, 1956 *Proceedings*.)

Captain Elmer W. Malanot.—It appears that, unfortunately, there was no one who actually shut down completely, or at least to a dead slow, the power of his engines, near the center of a cyclonic storm, so that he could have judged the wisdom of such action.

We had this opportunity again on this vessel on October 18, 1955, on a voyage from Saigon to Yokohama, when sixty miles south of Myako Shima, we ran into Typhoon Opal. We had no previous warn-

* From United States Naval Institute *Proceedings*—March 1956.

ing, when at 180600 I the barometer started to fall rapidly, and the wind from NNE increased to full gale force. The engine power was gradually reduced and, when the sea became confused at 181137 I, completely shut down to a dead slow. This vessel being powered with turbines, this was done merely to keep the engines warm. The vessel drifted in a southwesterly direction at the rate of three knots. At 1600 I, the wind was at its maximum of about 100 knots. Then the wind backed to NNW. The vessel was heading between 230 degrees and 265 degrees, and drifting in a southeasterly direction. The sea was mountainous and confused. The vessel pitched heavily and rolled to a maximum of 35 degrees.

Although this vessel was only ten miles from the center of this moderate typhoon, we had absolutely no damage whatsoever, while the ship had a full deck load of heavy Army trucks, lashed in the usual manner. And so again the only thing we did was to stop our engines.

At 182120 I, the force of wind and sea gradually moderated, and we proceeded to Yokohama.

This theory of evading damage by shutting down the power that primarily produces such damage, is so simple that, because of its simplicity, many people who have had experience in heavy weather, are so overawed by the gigantic force of nature, that they are reluctant to accept this simple fact.

I am including a letter I received from a Naval Architect, in which you might be interested, and you have my permission to reprint any of the above in your *Proceedings* if you so desire.

(Editor's Note: The following are excerpts from a letter Captain Malanot recently received.)

Recently I read your very enlightening article "Typhoon Doctrine." Your observations and recommendations are not only of value to seafaring men but also to Naval Architects.

Your observations and theory supporting your "Doctrine" are very convincing. Oftentimes observations do not seem to support a theoretical analysis; such divergences are passed over by the overworked proverb "practice and theory do not agree." In reality, when this is the case, there is something wrong with either one or the other.

I can well imagine that it is difficult to convince the crew that passive defense, as you put it, is the way to weather a typhoon. I am sure that you have considered the comments I wish to make.

If I may, I will briefly consider the stress aspect set up in machinery. Ship machinery is designed to withstand stresses due to torsion of the machine, shock stress due to load surges, inertia stresses due to rolling and pitching of the ship and others. Except for torsional stresses of the machine under consideration, the other stresses are difficult to evaluate since the sea forces inducing these stresses are difficult to evaluate and are impossible to control. These stresses are not always accumulative, but potentially there is the ever present chance that they can be. For heavy machinery, or for lightweight high-speed machinery in motion, inertia stresses due to rolling, especially for a "stiff ship," can attain tremendous values. In fact this stress can be of such proportions that structurally it can not be resisted. With stopped engines the torsional stresses due to rotation are zero, load surge stresses are zero, and pitching and rolling inertia stresses are lessened since minimum resistance is offered to the irresistible typhoon sea waves. Another type of machinery failure can also be avoided by your doctrine. Heavy rotating masses, like shafts, can have induced deflections by the forces mentioned to the extent that bearings will fail immediately. Also rolling sea suctions out of the water, loss of oil pressure and similar failures experienced in storms will result in major machinery derangements which in themselves can cause the loss of a ship. Rolling a boiler loose from its foundation is not uncommon. Plowing into a typhoon is just asking for such failures. Following such machinery derangements, uncontrollable engine room fires and similar disasters surely follow which spell doom for the ship. A dead plant or a minimum operative one while in a typhoon, should lessen the ever impending machinery failures mentioned.

I am sure that every Naval Architect would be relieved of much doubt if he knew that all ship Masters had your respect for the forces of the sea. Again, I want to assure you that your article made me realize that a designer does not have to do the impossible; namely, to build a ship to withstand almost irresistible forces.

Sincerely yours,

HOWARD A. PETERSON

DISSENT FROM THE MALANOT TYPHOON DOCTRINE*

COMMANDER W. J. RUHE, USN.—Malanot's theory that a ship can safely lie to in a trough of typhoon waves appears to be based on experience in broad-beamed types of ships, not with narrow-beamed, topside-heavy warships. Subsequent printed discussion on this theory interpreted Admiral Nimitz's letter as furthering the idea that ships might be given their head *in extremis*. However, on analysis, it appears that the Admiral's letter merely says that there are times when the skipper of a warship must leave formation by changing speed or course in order to save his ship. To this I agree. But where has Nimitz's letter recommended that a warship be allowed to lie to?

There is further evidence in Admiral Nimitz's letter which indicates that the worst thing for a warship is to go dead and lie to, thus losing its head and falling off into the trough. The *Monaghan* and the *Hull*, it can be presumed, lost all power and fell off into the trough. Since both were well fueled (75%) the finger cannot be placed on their being improperly ballasted. The *Dewey's* skipper, it is noted, attributed his survival to keeping way-on with full rudder. But the general tenor of all of the articles is that a warship skipper might benefit from Malanot's theory.

Let this be a dissenting vote to provide some form of sanity to the discussion. As a destroyer skipper I fear that just one other skipper will believe the theory and lose a ship.

Would a single naval constructor go along with Malanot's theory as it applied to ships like destroyers?

* From United States Naval Institute *Proceedings*—July 1956.

Appendix VIII : Extracts from *Typhoon*[*]

By Joseph Conrad

Chapter Two—Observing the steady fall of the barometer, Captain MacWhirr thought, "There's some dirty weather knocking about." This is precisely what he thought. He had had an experience of moderately dirty weather—the term dirty as applied to the weather implying only moderate discomfort to the seaman. Had he been informed by an indisputable authority that the end of the world was to be finally accomplished by a catastrophic disturbance of the atmosphere, he would have assimilated the information under the simple idea of dirty weather, and no other, because he had no experience of cataclysms, and belief does not necessarily imply comprehension. The wisdom of his county had pronounced by means of an Act of Parliament that before he could be considered as fit to take charge of a ship he should be able to answer certain simple questions on the subject of circular storms such as hurricanes, cyclones, typhoons; and apparently he had answered them, since he was now in command of the *Nan-Shan* in the China seas during the season of typhoons. But if he had answered he remembered nothing of it. He was, however, conscious of being made uncomfortable by the clammy heat. He came out on the bridge, and found no relief to this oppression. The air seemed thick. He gasped like a fish, and began to believe himself greatly out of sorts.

The *Nan-Shan* was ploughing a vanishing furrow upon the circle of the sea that had the surface and the shimmer of an undulating piece of gray silk. The sun, pale and without rays, poured down leaden heat in a strangely indecisive light, and the Chinamen were lying prostrate about the decks.

. . . The lurid sunshine cast faint and sickly shadows. The swell ran higher and swifter every moment, and the ship lurched heavily in the smooth, deep hollows of the sea.

"I wonder where that beastly swell comes from," said Jukes [Chief Mate] aloud, recovering himself after a stagger.

"North-east," grunted the literal MacWhirr, from his side of the bridge. "There's some dirty weather knocking about. Go and look at the glass."

. . . At its setting the sun had a diminished diameter and an expiring brown, rayless glow, as if millions of centuries elapsing since the morning had brought it near its end. A dense bank of cloud became visible to the northward; it had a sinister dark olive tint, and lay low and motionless upon the sea, resembling a solid obstacle in the path of the ship. She went floundering towards it like an exhausted creature driven to its death. The coppery twilight retired slowly, and the darkness brought out overhead a swarm of unsteady, big stars, that, as if blown upon, flickered exceedingly and seemed to hang very near the earth. At eight o'clock Jukes went into the chart-room to write up the ship's log.

. . . "Ship rolling heavily in a high cross swell," he began again, and commented to himself, "Heavily is no word for it." Then he wrote: "Sunset threatening, with a low bank of clouds to N. and E. Sky clear overhead."

Sprawling over the table with arrested pen, he glanced out of the door, and in that frame of his vision he saw all the stars flying upwards between the teak-wood jambs on a black sky. The whole lot took flight together and disappeared, leaving only a blackness flecked with white flashes, for the sea was as black as the sky and speckled with foam afar. The stars that had flown to the roll came back on the return swing of the ship, rushing downwards in their glittering multitude, not of fiery points, but enlarged to tiny discs brilliant with a clear wet sheen.

Jukes watched the flying big stars for a moment, and then wrote: "8 P.M. Swell increasing. Ship labouring and taking water on her decks. Battened down the coolies for the night. Barometer still falling." He paused, and thought to himself, "Perhaps nothing whatever'll come of it." And then he closed resolutely his entries: "Every appearance of a typhoon coming on." . . .

Chapter Three—. . . The strong wind swept at him [Captain MacWhirr] out of a vast obscurity; he felt under his feet the uneasiness of his ship, and he could not even discern the shadow of her shape. He wished it were not so; and very still he waited, feeling stricken by a blind man's helplessness.

To be silent was natural to him, dark or shine. Jukes, at his elbow, made himself heard yelling cheerily in the gusts, "We must have got the worst of it at once, sir." A faint burst of lightning quivered all round, as if flashed into a cavern—into a black and secret chamber of the sea, with a floor of foaming crests.

It unveiled for a sinister, fluttering moment a ragged mass of clouds hanging low, the lurch of the long outlines of the ship, the black figures of men caught on the bridge, heads forward, as if petrified in the act of butting. The darkness palpitated down upon all this, and then the real thing came at last.

It was something formidable and swift, like the sudden smashing of a vial of wrath. It seemed to explode all round the ship with an overpowering concussion and a rush of great waters, as if an immense dam had been blown up to windward. In an instant the men lost touch of each other. This is the disintegrating power of a great wind: it isolates one from one's kind.

. . . The rain poured on him [Jukes], flowed, drove in sheets. He breathed in gasps; and sometimes the water he swallowed was fresh and sometimes it was salt. For the most part he kept his eyes shut tight, as if suspecting his sight might be destroyed in the immense flurry of the elements. When he ventured to blink hastily, he derived some moral support from the green gleam of the starboard light shining feebly upon the flight of rain and sprays. He was actually looking at it when its ray fell upon the uprearing sea which put it out. He saw the head of the wave topple over, adding the mite of its crash to the tremendous uproar raging around him, and almost at the same instant the stanchion was wrenched away from his embracing arms. After a crushing thump on his back he found himself suddenly afloat and borne upwards. His first irresistible notion was that the whole China Sea had climbed on the bridge. Then, more sanely, he concluded himself gone overboard. All the time he was being tossed, flung, and rolled in great volumes of water, he kept on repeating mentally, with the utmost precipitation, the words: "My God! My God! My God!"

. . . The motion of the ship was extravagant. Her lurches had an appalling helplessness: she pitched as if taking a header into a void, and seemed to find a wall to hit every time. When she rolled she fell on her side headlong, and she would be righted back by such a demolishing blow that Jukes felt her reeling as a clubbed man reels before he collapses. The gale howled and scuffled about gigantically in the darkness, as though the entire world were one black gully. At certain moments the air streamed against the ship as if sucked through a tunnel with a concentrated solid force of impact that seemed to lift her clean out of the water and keep her up for an instant with only a quiver running through her from end to end. And then she would begin her tumbling again as if dropped back into a boiling cauldron. Jukes tried hard to compose his mind and judge things coolly.

The sea, flattened down in the heavier gusts, would uprise and overwhelm both ends of the *Nan-Shan* in snowy rushes of foam, expanding wide, beyond both rails, into the night. And on this dazzling sheet, spread under the blackness of the clouds and emitting a bluish glow, Captain MacWhirr could catch a desolate glimpse of a few tiny specks black as ebony, the tops of the hatches, the battened companions, the heads of the covered winches, the foot of a mast. This was all he could see of his ship. Her middle structure, covered by the bridge which bore him, his mate, the closed wheelhouse where a man was steering shut up with the fear of being swept overboard together with the whole thing in one great crash—her middle structure was like a half-tide rock awash upon a coast. It was like an outlying rock with the water boiling up, streaming over, pouring off, beating round—like a rock in the surf to which shipwrecked people cling before they let go—only it rose, it sank, it rolled continuously, without respite and rest, like a rock that should have miraculously struck adrift from a coast and gone wallowing upon the sea.

The *Nan-Shan* was being looted by a storm with a senseless, destructive fury: trysails torn out of the extra gaskets, double-lashed awnings blown away, bridge swept clean, weather-cloths burst, rails twisted, light-screens smashed—and two of the boats had gone already. They had gone unheard and unseen, melting, as it were, in the shock and smother of the wave. It was only later, when upon the white flash of another high sea hurling itself amidships, Jukes had a vision of two pairs of davits leaping black and empty out of the solid blackness, with one overhauled fall flying and an iron-bound block capering in the air, that he became aware of what had happened within about three yards of his back.

. . . They [Captain MacWhirr and Jukes] held hard. An outburst of unchained fury, a vicious rush of the wind absolutely steadied the ship; she rocked only, quick and light like a child's cradle, for a terrific moment of suspense, while the whole atmosphere, as is seemed, streamed furiously past her, roaring away from the tenebrous earth.

It suffocated them, and with eyes shut they tightened their grasp. What from the magnitude of the shock might have been a column of water running upright in the dark, butted against the ship, broke short, and fell on her bridge, crushingly, from on high, with a dead burying weight.

A flying fragment of that collapse . . . enveloped them in one swirl from their feet over their heads, filling violently their ears, mouths and nostrils with salt water. It knocked out their legs, wrenched in haste at their arms, seethed away swiftly under their chins; and opening their eyes, they saw the piled-up masses of foam dashing to and fro amongst what looked like the fragments of a ship. She had given way as if driven straight in. Their panting hearts yielded, too, before the tremendous blow; and all at once she sprang up again to her desperate plunging, as if trying to scramble out from under the ruins.

The seas in the dark seemed to rush from all sides to keep her back where she might perish. There was hate in the way she was handled, and a ferocity in the blows that fell. She was like a living creature thrown to the rage of a mob: hustled terribly, struck at, borne up, flung down, leaped upon. Captain MacWhirr and Jukes kept hold of each other, deafened by the noise, gagged by the wind; and the great physical tumult beating about their bodies, brought, like an unbridled display of passion, a profound trouble to their souls. One of those wild and appalling shrieks that are heard at times passing mysteriously overhead in the steady roar of a hurricane, swooped, as if borne on wings, upon the ship, and Jukes tried to outscream it.

"Will she live through this?"

The cry was wrenched out of his breast. It was as unintentional as the birth of a thought in the head, and he heard nothing of it himself. It all became extinct at once—thought, intention, effort—and of his cry the inaudible vibration added to the tempest waves of the air.

He expected nothing from it. Nothing at all. For indeed what answer could be made? But after a while he heard with amazement the frail and resisting voice in his ear, the dwarf sound, unconquered in the giant tumult.

"She may!" . . .

Chapter Five—. . . Nobody—not even Captain MacWhirr, who alone on deck had caught sight of a white line of foam coming on at such a height that he couldn't believe his eyes—nobody was to know the steepness of the sea and the awful depth of the hollow the hurricane had scooped out behind the running wall of water.

It raced to meet the ship, and, with a pause, as of girding the loins, the *Nan-Shan* lifted her bows and leaped. The flames in all the lamps sank, darkening the engine-room. One went out. With a tearing crash and a swirling, raving tumult, tons of water fell upon the deck, as though the ship had darted under the foot of a cataract.

Down there they looked at each other, stunned.

"Swept from end to end, by God!" bawled Jukes.

She dipped into the hollow straight down, as if going over the edge of the world. The engine-room toppled forward menacingly, like the inside of a tower nodding in an earthquake. An awful racket, of iron things falling, came from the stokehold. She hung on this appalling slant long enough for Beale [Third Engineer] to drop on his hands and knees and begin to crawl as if he meant to fly on all fours out of the engine-room, and for Mr. Rout [Chief Engineer] to turn his head slowly, rigid, cavernous, with the lower jaw dropping. Jukes had shut his eyes, and his face in a moment became hopelessly blank and gentle, like the face of a blind man.

At last she rose slowly, staggering, as if she had to lift a mountain with her bows.

Mr. Rout shut his mouth; Jukes blinked; and little Beale stood up hastily.

"Another one just like this, and that's the last of her," cried the chief.

. . . There was no wind, not a breath, except the faint currents created by the lurches of the ship. The smoke tossed out of the funnel was settling down upon her deck. He [Jukes] breathed it as he passed forward. He felt the deliberate throb of the engines, and heard small sounds that seemed to have survived the great uproar: the knocking of broken fittings, the rapid tumbling of some piece of wreckage on the bridge. He perceived dimly the squat shape of his captain holding on to a twisted bridge-rail, motionless and swaying as if rooted to the planks. The unexpected stillness of the air oppressed Jukes.

"We have done it, sir," he gasped.

. . . After the whisper of their shouts, their ordinary tones, so distinct, rang out very loud to their ears in the amazing stillness of the air. It seemed to them they were talking in a dark and echoing vault.

Through a jagged aperture in the dome of clouds the light of a few stars fell upon the black sea, rising and falling confusedly.

Sometimes the head of a watery cone would topple on board and mingle with the rolling flurry of foam on the swamped deck; and the *Nan-Shan* wallowed heavily at the bottom of a circular cistern of clouds. This ring of dense vapours, gyrating madly round the calm of the centre, encompassed the ship like a motionless and unbroken wall of an aspect inconceivably sinister. Within, the sea, as if agitated by an internal commotion, leaped in peaked mounds that jostled each other, slapping heavily against her sides; and a low moaning sound, the infinite plaint of the storm's fury, came from beyond the limits of the menacing calm. Captain MacWhirr remained silent, and Jukes' ready ear caught suddenly the faint, long-drawn roar of some immense wave rushing unseen under the thick blackness, which made the appalling boundary of his vision.

. . . The worst was to come, then—and if the books were right this worst would be very bad. The experience of the last six hours had enlarged his [Captain MacWhirr's] conception of what heavy weather could be like. "It'll be terrific," he pronounced, mentally.

. . . He [Captain MacWhirr] released it [box of matches] at last, and letting himself fall on the settee, listened for the first sounds of returning wind.

Not yet. He heard only the wash of water, the heavy splashes, the dull shocks of the confused seas boarding his ship from all sides. She would never have a chance to clear her decks.

But the quietude of the air was startlingly tense and unsafe, like a slender hair holding a sword suspended over his head. By this awful pause the storm penetrated the defences of the man and unsealed his lips. He spoke out in the solitude and the pitch darkness of the cabin, as if addressing another being awakened within his breast.

"I shouldn't like to lose her," he said half aloud.

. . . "It will come very sudden," said Captain MacWhirr, "and from over there, I fancy. God only knows though. These books are only good to muddle your head and make you jumpy. It will be bad, and there's an end. If we only can steam her round in time to meet it. . . ."

A minute passed. Some of the stars winked rapidly and vanished.

. . . A hollow echoing noise, like that of a shout rolling in a rocky chasm, approached the ship and went away again. The last star, blurred, enlarged, as if returning to the fiery mist of its beginning, struggled with the colossal depth of blackness hanging over the ship—and went out.

"Now for it!" muttered Captain MacWhirr. "Mr. Jukes."

"Here, sir."

The two men were growing indistinct to each other.

"We must trust her to go through it and come out on the other side. That's plain and straight. There's no room for Captain Wilson's storm-strategy here."

"No, sir."

"She will be smothered and swept again for hours," mumbled the Captain. "There's not much left by this time above deck for the sea to take away—unless you or me."

"Both, sir," whispered Jukes, breathlessly.

"You are always meeting trouble half way, Jukes," Captain MacWhirr remonstrated quaintly.

"Though it's a fact that the second mate is no good. D'ye hear, Mr. Jukes? You would be left alone if. . . ."

Captain MacWhirr interrupted himself, and Jukes, glancing on all sides, remained silent.

"Don't you be put out by anything," the Captain continued, mumbling rather fast. "Keep her facing it. They may say what they like, but the heaviest seas run with the wind. Facing it—always facing it—that's the way to get through. You are a young sailor. Face it. That's enough for any man. Keep a cool head."

Bibliography

BOOKS

CLINE, I. M. *Tropical Cyclones.* New York: MacMillan Company, 1926.

DOUGLAS, MARJORY STONEMAN. *Hurricane.* New York: Rinehart & Co., 1958.

DUNN, G. E. "Tropical Cyclones," *Compendium of Meteorology.* Boston: American Meteorological Society, 1951.

——— and MILLER, B. I. *Atlantic Hurricanes.* Baton Rouge: Louisiana State University Press, 1964.

HUBERT, L. F., and CLARK, G. B. *The Hurricane Surge.* Washington: U. S. Weather Bureau, 1955.

MILLER, B. I. *The Three Dimensional Wind Structure Around a Tropical Cyclone.* National Hurricane Research Project, Report No. 15, 1958.

NAMIAS, J. *Forms of the General Circulation Related to Hurricane Genesis and Path.* Proceedings, Technical Conference on Hurricanes, Miami Beach, Fla., November 20-22, 1958.

RIEHL, H. "Aerology of Tropical Storms," *Compendium of Meteorology.* Boston: American Meteorological Society, 1951.

———. *Tropical Meteorology.* New York: McGraw-Hill Book Company, 1954.

———. *Comments on the Formation of Hurricanes.* Proceedings, Technical Conference on Hurricanes, Miami Beach, Fla., November 20-22, 1958.

TANNEHILL, I. R. *Hurricanes, Their Nature and History.* Princeton: Princeton University Press, 1938.

———. *Hurricane Hunters.* New York: Dodd, Mead, and Co., 1954.

U. S. NAVY, BUREAU OF AERONAUTICS. *Marine Climatic Atlas of the World, North Atlantic Ocean* (Vol. I). 1955.

———. *Intensification of Tropical Cyclones, Atlantic and Pacific Areas.* Fourth Research Project, Project AROWA, Norfolk, Va., 1956.

U. S. WEATHER BUREAU. *Atlas of Climatic Charts of the Oceans.* Washington, 1938.

WEEMS, JOHN E. *A Weekend in September.* New York: Henry Holt & Co., 1957.

WILLETT, H. C. *A Study of Tropical Hurricanes Along the Atlantic and Gulf Coasts of the United States.* New York: Inter-regional Insurance, 1955.

PERIODICALS

CLAPP, P. F., and TISDALE, C. F. "Origin and Paths of Hurricanes and Tropical Storms Related to Certain Physical Parameters at the Air-Sea Interface," *Journal of Applied Meteorology,* 2 (1963), 358-67.

COLON, J. A. "A Study of Hurricane Tracks for Forecasting Purposes," U. S. Weather Bureau *Monthly Weather Review,* 81 (1953), 53-66.

———. "On the Formation of Hurricane Alice, 1955, with Notes on Other Cold-Season Tropical Storms," U. S. Weather Bureau *Monthly Weather Review,* 84 (1956), 1-14.

———. and NIGHTINGALE, W. R. "Development of Tropical Cyclones in Relation to Circulation Patterns at the 200-Millibar Level," U. S. Weather Bureau *Monthly Weather Review* (1963), 329-36.

CONNER, W. C., KRAFT, R. H., and HARRIS, D. L. "Empirical Methods for Forecasting the

Maximum Storm Tide Due to Hurricanes and Other Tropical Storms," U. S. Weather Bureau *Monthly Weather Review,* 85 (1957), 113-16.

DUNN, G. E. "Cyclogenesis in the Tropical Atlantic," American Meteorological Society *Bulletin,* 21 (1940), 215-29.

———. "Areas of Hurricane Development," U. S. Weather Bureau *Monthly Weather Review,* 84 (1956), 57-61.

FLETCHER, R. D. "Computation of Maximum Winds in Hurricanes," American Meteorological Society *Bulletin,* 36 (1955), 246-50.

GENTRY, R. C. "Wind Velocities During Hurricanes," Paper No. 2731, American Society of Civil Engineers, 120 (1955), 169.

HAGGARD, W. H., and CRY, A. W. "North Atlantic Tropical Cyclone Activity," U. S. Weather Bureau *Monthly Weather Review,* (1962), 341-49.

HOOVER, R. A. "Empirical Relationships of the Central Pressures in Hurricanes to the Maximum Surge and Storm Tide," U. S. Weather Bureau *Monthly Weather Review,* 85 (1957), 167-74.

JORDAN, C. L., and TE-CHUN HO. "Variations in the Annual Frequency of Tropical Cyclones," U. S. Weather Bureau *Monthly Weather Review* (1962), 157-64.

KNOX, J. L. "The Storm Hazel," American Meteorological Society *Bulletin,* 36 (1955), 239-46.

LABAN, J. "Sequence of Weather Events, Hurricane Janet, September 26-27, 1955. Swan Island, West Indies." *Weatherwise,* 9 (1955), 121-24.

LANDERS, H. "On the Formation and Development of Tropical Cyclones," *Journal of Applied Meteorology,* 2 (1963), 206-18.

MALKUS, J. "On the Structure and Maintenance of the Mature Hurricane Eye," *Journal of Applied Meteorology,* 15 (1958), 337-49.

MILLER, B. I. "On the Maximum Intensity of Hurricanes," *Journal of Applied Meteorology,* 15 (1958), 185-95.

———. "Rainfall Rates in Florida Hurricanes," U. S. Weather Bureau *Monthly Weather Review,* 86 (1958), 258-64.

NAMIAS, J. "Tropical Cyclones Related to the General Circulation," New York Academy of Science, 1955.

RIEHL, H., HAGGARD, W. H., and SANBORN, R. W. "On the Prediction of 24-hour Hurricane Movement," *Journal of Applied Meteorology,* 13 (1956), 415-20.

SIMPSON, R. H. "On the Movement of Tropical Cyclones," American Geophysical Union, 27 (1946), 641-45.

———. "Exploring the Eye of Typhoon Marge, 1951," American Meteorological Society *Bulletin,* 33 (1952), 286-98.

———. "Hurricanes," *Scientific American,* June, 1954.

Index

Part I–Hurricanes

Index

Part II–Typhoons

The text of this book is set in Linotype Caledonia. It is twelve point with two points of leading. The section titles are eighteen point Caledonia Bold.

The book is printed on White Butte des Morts Offset paper. The cover is Lexotone No. 22598, charcoal grey, kid grain.

Editorial production by Dorothy Williams.

Design by Gerard Valerio.

The book was composed, printed, and bound by George Banta Company, Incorporated, Menasha, Wisconsin.